MODELLING UNCERTAINTY IN FLOOD FORECASTING SYSTEMS

T0362088

Modelling Uncertainty in Flood Forecasting Systems

DISSERTATION

Submitted in fulfilment of the requirements of
the Board for Doctorates of Delft University of Technology and of the
Academic Board of UNESCO-IHE Institute for Water Education
for the Degree of DOCTOR
to be defended in public
on Monday, 24 May 2004 at 10:30 hours
in Delft, The Netherlands

by

Shreedhar MASKEY

*Civil Engineer, Master of Science in Hydroinformatics with Distinction
born in Charikot, Dolkha district, Nepal*

Taylor & Francis
Taylor & Francis Group

LONDON AND NEW YORK

This dissertation has been approved by the promoter
Prof. dr. R.K. Price, TU Delft / UNESCO-IHE Delft, The Netherlands

Members of the Awarding Committee:

Chairman	Rector Magnificus TU Delft, The Netherlands
Co-chairman	Director UNESCO-IHE Delft, The Netherlands
Prof. dr. R.K. Price	TU Delft / UNESCO-IHE, The Netherlands, promoter
Prof. dr. A.W. Heemink	TU Delft, The Netherlands
Prof. dr. J.P. O'Kane	University College Cork, Ireland
Prof. dr.-ing. A. Bardossy	University of Stuttgart, Germany
Prof. dr. M.J. Hall	UNESCO-IHE Delft, The Netherlands
Dr. V. Guinot	University of Montpellier 2, France
Prof. dr. H.H.G. Savenije	TU Delft, The Netherlands, reserved member

Copyright © 2004 Taylor & Francis Group plc, London, UK

All rights reserved. No part of this publication or the information contained herein may be reproduced, stored in a retrieval system, or transmitted in any form or by any means, electronic, mechanical, by photocopying, recording or otherwise, without written prior permission from the publisher.

Although all care is taken to ensure the integrity and quality of this publication and the information herein, no responsibility is assumed by the publishers nor the authors for any damage to property or persons as a result of operation or use of this publication and/or the information contained herein.

Published by Taylor & Francis
2 Park Square, Milton Park, Abingdon, Oxon, OX14 4RN
270 Madison Ave, New York NY 10016

Transferred to Digital Printing 2007

ISBN 90 5809 694 7 (Taylor & Francis Group)

Publisher's Note
The publisher has gone to great lengths to ensure the quality of this reprint but points out that some imperfections in the original may be apparent

Summary

MODELLING UNCERTAINTY IN FLOOD FORECASTING SYSTEMS

Uncertainty is a common factor of everyday life. In almost all circumstances we find ourselves in a state of uncertainty. The subject of this research is about uncertainty in flood forecasting systems. Like all natural hazards flooding is a complex and inherently uncertain phenomenon. The uncertainty in flood forecasting comes from the uncertainty in the precipitation forecast and other inputs, the model parameters, model structure and so forth. Despite increasing advances in the development of flood forecasting models and techniques, the uncertainty in actual flood forecasts remains unavoidable. It is therefore important that the existence of uncertainty in flood forecasts be admitted and properly appraised. Hiding uncertainty may create the illusion of certainty, the consequences of which can be very significant. The major benefits of estimating uncertainty in flood forecasting are that it provides a rational basis for flood warning (risk-based warning) and it offers potential by providing economic benefits from flood forecasting and warning systems. Even with the lack of risk-based flood warning procedures, quantifying uncertainty provides additional information about the forecasts and helps decision makers to use their own judgement more appropriately.

The aim of this research is to develop a framework and tools and techniques for modelling uncertainty in flood forecasting systems. This research applies uncertainty assessment methods based on probability theory and fuzzy set theory to the problem of flood forecasting. The outcomes of this research can be summarised as follows:

1. Review of different types of flood forecasting models and their appraisal with respect to uncertainty assessment.

2. Review of uncertainty representation and modelling theories and methods, particularly in regard to flood forecasting.

3. Development of a methodology using temporal disaggregation for uncertainty assessment in model outputs due to the uncertainty in time series inputs.

4. Development of an Improved First-Order Second Moment (IFOSM) method for uncertainty assessment using a second-order reconstruction of the model function.

5. Development of a qualitative uncertainty scale using the best-case and worst-case scenarios for the interpretation of the uncertainty analysis results generated by the expert judgement-based qualitative method.

6. Exploration of hybrid techniques of uncertainty modelling and probability-possibility transformations.

Different types of models for flood forecasting are reviewed. Broadly they can be classified as (i) physically-based, (ii) conceptual and empirical, and (iii) data-driven. Uncertainty analysis is a well accepted procedure in the first two categories of models, although many operational flood forecasting systems may not have components for uncertainty analysis. Normally, increasing the complexity of the model decreases the uncertainty in the model output due to the model structure. In this sense, model uncertainty should in principle decrease as we move from empirical to conceptual to physically based models. But model uncertainty is only a part of the total uncertainty. If there exists a large uncertainty in the inputs and if the parameters cannot be estimated with high precision, using complex models does not guarantee less uncertain model results. Some of the data-driven techniques, such as fuzzy rule-based systems and fuzzy regression, work with imprecise data and implicitly incorporate the uncertainty concept in modelling. These models however, do not have the flexibility of using uncertainty methods based on other theories (e.g. the most popular probability theory). In the case of the model based on artificial neural network (ANN) (the most popular so far of the data-driven techniques), model performance is, by and large, expressed in a form based on the difference between the observed and model predicted results using measures such as Root Mean Square Error (RMSE). More study is required for the application of the more commonly used uncertainty analysis techniques, such as those based on probability and fuzzy set theories, to ANN-based models.

The literature review on uncertainty suggests that probability theory and fuzzy set theory (including fuzzy measures and possibility theory) are the two most widely used theories for uncertainty representation. Probability theory assumes uncertainty mainly due to randomness, whereas fuzzy set theory assumes it due to vagueness (or fuzziness) and imprecision. In flood forecasting, however, the application of theories other than probability has so far been insignificant. It is argued that these two theories must be treated as complementary rather than competitive. In this research the Monte Carlo method and the First-Order Second Moment (FOSM) method are used as the standard techniques for uncertainty propagation in the probabilistic approach. At the same time, this research extends the application of fuzzy set theory for uncertainty modelling in flood forecasting. In particular two methods based on fuzzy set theory are explored: (i) the fuzzy Extension Principle, and (ii) the expert judgement based qualitative method.

The development of the methodology using temporal disaggregation for uncertainty assessment in model results due to the uncertainty in time series inputs is an important contribution of this research. This methodology can be implemented in the frameworks of both the Monte Carlo method and the fuzzy Extension Principle, and can be applied to any type of deterministic rainfall-runoff model. This methodology explicitly requires the uncertainty in the time series inputs to be represented by probability distributions for the probabilistic approach or by membership functions for the fuzzy approach. The latter approach is particularly useful when the precipitation forecasts are nonprobabilistic.

The fact that the popular FOSM method uses the linearisation of the function gives rise to some limitations of the method. As part of this research, an Improved FOSM method is proposed using the second-degree reconstruction of the function to be modelled. The IFOSM method retains the simplicity and the smaller computational requirement of the FOSM method and it has a particular advantage when the average value of the input variable corresponds to maximum/minimum of model function or to values in regions where the slope of the function is very mild compared to the effects of curvature (non-linearity).

This research also demonstrates the application of a qualitative method of uncertainty assessment in flood forecasting. The qualitative method is based on expert judgement and fuzzy set theory. The advantage of this method is that it allows the incorporation of uncertainty due to all recognisable sources without a significant increase in the computation time. This method has the same mathematical structure as that of the FOSM method but the evaluations of the *quality* and *importance* (similar to the variance and sensitivity, respectively, of the FOSM method) are fully based on expert judgement. Consequently it is a very approximate yet more holistic method. One of the issues in using this method concerns the interpretation of the results. In the course of this research, a Qualitative Uncertainty Scale was developed on which the results from the expert judgement-based method can be classified qualitatively using linguistic variables such as small uncertainty, moderate uncertainty, large uncertainty, etc. The derivation of the uncertainty scale is based on the concept of best-case and worst-case scenarios.

This research also investigates the possibility of a hybrid technique of uncertainty modelling, whereby the combined application of both probabilistic and possibilistic or fuzzy approaches are used. There are at least two types of situation where the hybrid technique can be useful, which are referred in this thesis as Type I and Type II Problems. The Type I Problem is the situation where an uncertain variable possesses components from randomness and fuzziness. The Type II Problem refers to the situation where there exist two sets of uncertainty parameters: one represented by probability distributions and the other represented by possibility distributions or membership functions. Whereas the concepts, such as fuzzy probability and fuzzy-random variable can be used to deal with the Type I Problem, a solution to the Type II Problem is also possible if an acceptable transformation between probability and possibility can be established.

The major obstacle in modelling uncertainty of the Type II Problem comes from the differences of operation between random-random (R-R) and fuzzy-fuzzy (F-F) variables. Part of this research is also devoted to the exploration of the differences and similarities in operation between R-R and F-F variables, and to investigate the derivation of a probability-possibility (or fuzzy) transformation that takes these differences into account. This research shows that the arithmetic operations between two fuzzy variables by the fuzzy Extension Principle using a method called α-cut are similar to corresponding operations between two functionally dependent random variables for some specific conditions. It also provides an alternative method for the evaluation of the Extension Principle for a monotonic function without using the α-cut method. Although the direct implication of this finding is very limited, it is certainly an important basis for further research in probability-possibility transformations and hybrid approaches in uncertainty modelling.

The application examples used in this research are flood forecasting models for the Klodzko catchment in Poland and the Loire River in France. The uncertainty assessment methodology using temporal disaggregation is applied to the Klodzko model to assess the uncertainty in the forecasted floods due to the uncertainty in the precipitation time series. It is implemented in the framework of the fuzzy Extension Principle and assisted by genetic algorithms (GAs) for the determination of the minimum and maximum of the model function. The results are illustrated using two versions of GAs, namely a normal (or conventional) GA and a micro GA. The Loire model is used to apply the FOSM, IFOSM and expert judgement-based qualitative methods for uncertainty assessment in the model forecasts due to the uncertainty in different parameters. In this application the Monte Carlo method is used as a standard method to compare the results from the FOSM and IFOSM methods.

Shreedhar Maskey

Delft, 2004

Table of Contents

Summary v

Table of Contents ix

1. INTRODUCTION 1

1.1 Flood and flood hazard 1

1.1.1 International dimension of the consequences of flooding 2

1.1.2 A reminder of a few recent floods 3

1.2 Flood management options 4

1.2.1 Alternative strategies of flood alleviation 4

1.2.2 Flood forecasting, warning and response systems 5

1.3 Subject of the present research: uncertainty in flood forecasting and warning . 7

1.3.1 Uncertainty in flood forecasting and warning 7

1.3.2 Benefits of uncertainty quantification in flood forecasting 8

1.3.3 Aim and objectives of the present research 9

1.3.4 Application examples 9

1.4 Structure of the thesis 9

2. FLOOD FORECASTING MODELS AND UNCERTAINY
   REPRESENTATION 11

2.1 Types of flood forecasting models 11

2.1.1 Physically-based models 12

2.1.2 Conceptual and empirical models 12

2.1.3 Data-driven models 13

2.1.4 Flood forecasting models vis-à-vis uncertainty analysis 14

2.2 Uncertainty types and sources 15

2.2.1 Types of uncertainty 16

2.2.2 Sources of uncertainty in flood forecasting 17

2.3 Theories of uncertainty representation 17

2.3.1 Probability theory 18

2.3.2 Fuzzy set theory 21

2.3.3 Possibility theory and fuzzy measures 24

3. **EXISTING MATHEMATICAL METHODS FOR UNCERTAINTY ASSESSMENT** 27

    3.1 Probability theory-based methods..27

        3.1.1 Sampling method: Monte Carlo simulation 28

        3.1.2 Approximation method: FOSM 29

    3.2 Fuzzy set theory-based methods 31

        3.2.1 Fuzzy Extension Principle-based method 31

        3.2.2 Expert judgement-based qualitative method..35

    3.3 Bayesian theory-based uncertainty assessment methods used in flood forecasting..38

        3.3.1 Bayesian forecasting system (BFS) 39

        3.3.2 Generalised likelihood uncertainty estimation (GLUE) 39

    3.4 Hybrid techniques in uncertainty modelling.......................................40

        3.4.1 The concept of fuzzy probability 41

        3.4.2 The concept of fuzzy-random variable 41

    3.5 Methods for probability–possibility transformation 42

        3.5.1 Transformation by simple normalisation 43

        3.5.2 Transformation by principle of uncertainty invariance.........................44

    3.6 Discussion: probability– and fuzzy set theory–based methods.....................47

        3.6.1 Monte Carlo simulation and Extension Principle 48

        3.6.2 FOSM and expert judgement-based qualitative method.........................48

4. **CONTRIBUTION OF PRESENT RESEARCH TO UNCERTAINTY ASSESSMENT METHODS**..51

    4.1 Uncertainty assessment methodology using temporal disaggregation 51

        4.1.1 Principle of disaggregation 53

        4.1.2 Spatial variations of temporal distribution..55

        4.1.3 Uncertainty in the input quantity..55

        4.1.4 Algorithms for the fuzzy approach 56

        4.1.5 Algorithm for the Monte Carlo approach..57

        4.1.6 Generation of the pattern (disaggregation) coefficients.........................58

    4.2 Improved first–order second moment method.....................................61

        4.2.1 Practical implementation of FOSM method 62

        4.2.2 Problems attached to FOSM method 63

4.2.3 Principle of the improved method..64

4.2.4 Mathematical derivation..65

4.3 Qualitative scales for uncertainty interpretation 67

4.3.1 Derivation of Qualitative Uncertainty Scales 68

4.3.2 Presentation of uncertainty in Qualitative Uncertainty Scale 69

4.4 Towards hybrid techniques of modelling uncertainty...........................71

4.4.1 Type I and Type II Problems 71

4.4.2 Operations on random and fuzzy variables............................72

4.4.3 Probability–fuzzy transformations 77

4.4.4 Application example 79

4.4.5 Concluding remarks 82

5. APPLICATION: FLOOD FORECASTING MODEL FOR KLODZKO CATCHMENT (POLAND)..85

5.1 Klodzko catchment flood forecasitng model 85

5.1.1 Description of the model..86

5.1.2 Methods used to model rainfall-runoff processes 88

5.1.3 Uncertainty due to precipitation.......................................90

5.2 Implementation of the methodology for uncertainty assessment due to precipitation 91

5.2.1 Precipitation time series reconstruction using temporal disaggregation for uncertainty assessment 91

5.2.2 Precipitation uncertainty represented by a membership function 92

5.2.3 Algorithm for the propagation of uncertainty 94

5.2.4 Simplification of the methodology.....................................95

5.3 Genetic algorithms for minimum and maximum determination...................96

5.3.1 Principles of genetic algorithms.......................................96

5.3.2 GA versions used 98

5.4 Application and results..100

5.4.1 Results with 3 subperiods...102

5.4.2 Results with 3 and 6 subperiods......................................106

5.4.3 Results by normal GA and micro GA: a comparison 109

5.5 Conclusions and discussion 112

6. APPLICATION: FLOOD FORECASTING MODEL FOR LOIRE RIVER (FRANCE) 115

6.1 Loire River flood forecasting model 115

    6.1.1 Description of the model 118

    6.1.2 Sources of uncertainty 119

6.2 Uncertainty analysis using the FOSM method 121

    6.2.1 Description of data 121

    6.2.2 Results of analysis 122

    6.2.3 Conclusions 126

6.3 Application of the improved FOSM method 127

    6.3.1 Description of data 127

    6.3.2 Results of the analysis 128

    6.3.3 Conclusions 132

6.4 Uncertainty analysis using qualitative method 133

    6.4.1 Expert evaluation 133

    6.4.2 Analysis procedure and results 135

    6.4.3 Conclusions 137

6.5 Conclusions and discussion 138

7. CONCLUSIONS AND RECOMMENDATIONS 139

7.1 Conclusions 139

    7.1.1 Uncertainty in flood forecasting systems 139

    7.1.2 Theories and methods for modelling uncertainty 140

    7.1.3 Disaggregation of time series inputs for uncertainty assessment 140

    7.1.4 Use of genetic algorithms with fuzzy Extension Principle 141

    7.1.5 FOSM and Improved FOSM methods 141

    7.1.6 Expert judgement-based method and Qualitative Uncertainty Scales .. 142

    7.1.7 Probability-possibility transformation and hybrid technique for uncertainty modelling 142

7.2 Recommendations for further research and development 143

    7.2.1 Uncertainty assessment should be an integral component of flood forecasting systems 143

    7.2.2 Probabilistic and possibilistic approaches should be considered as complementary 143

    7.2.3 Towards hybrid techniques of uncertainty modelling 143

7.2.4 Towards uncertainty and risk-based flood forecasting and warning systems 144

7.2.5 Uncertainty assessment in data-driven modelling 144

I. FUZZY SETS, FUZZY ARITHMETIC AND DEFUZZIFICATION 147

I.1 Definitions on fuzzy sets 147

I.2 Fuzzy arithmetic 149

I.3 Defuzzification methods 151

Abbreviations 153

Notations 155

References 157

Samenvatting 169

Acknowledgements 175

About the Author 177

Chapter 1

INTRODUCTION

Summary of Chapter 1

The objective of this chapter is obviously to introduce the subject of this research, which is uncertainty in flood forecasting and warning systems. It starts with defining different types of flood and gives a brief introduction to the international dimension of the consequences of flood hazard. A few recent disastrous flood events are also presented. Secondly, different options of flood management are described with a more detailed introduction to flood forecasting, warning and response systems. Next, uncertainty issues in flood forecasting and warning are discussed, potential benefits of uncertainty assessment in flood forecasting are presented, the objectives of the research are set forth and two application examples are introduced. The structure of this thesis is outlined at the end of this chapter.

1.1 Flood and flood hazard

A flood can be defined as the inundation of a normally dry area caused by an increased water level in a watercourse or in a body of water. Broadly, flooding can be classified as *river flooding*, *coastal flooding* and *urban flooding*. This study is limited to the floods in rivers, which can be characterised by the sudden or gradual increase in water levels, normally beyond the bank level of the river watercourse and subsequently inundating its surroundings. From the nature of its occurrence, floods in rivers can be further distinguished as *flash flooding* and simply *river flooding*. Flash flooding is characterised by a sudden and massive increase of water quantity (e.g., caused by short intense bursts of rainfall, commonly from thunderstorms) often combined with mud or debris flows, leaving endangered communities only limited time to respond. River flooding, on the other hand, is marked by relatively slowly rising water levels of main rivers and a gradual inundation of floodplains due to generally continuous long duration rainfall. The former imposes a different and indeed more daunting challenge to the alertness of flood prone communities, whereas the larger scope and longer duration of the latter constitutes a major challenge to the scale and endurance of disaster management arrangements (Rosenthal et al., 1998). Various definitions of a flood and a comprehensive description of different types of flooding are presented by Ward (1978).

1.1.1 International dimension of the consequences of flooding

Floods remain one of the most frequent and devastating natural hazards worldwide. Internationally, floods pose a widely distributed natural risk to life, whereas other natural hazards such as avalanches, landslides, earthquakes and volcanic activities are more local or regional in their distribution (Samuels, 1999). Floods also cause impacts on society that go beyond economical cost and facilities, including impacts such as family and community disruptions, dislocation, injuries and unemployment. Figure 1.1 shows the loss of human life and the number of people affected due to major natural disasters in the world during 1966 to 1990 (see Fattorelli et al., 1999). During this period, the number of people affected by floods far exceeded the number affected by all other major disasters combined.

(a) Human Lives Lost

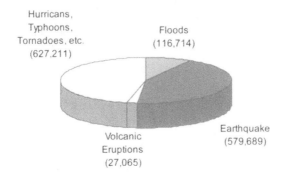

(b) People Affected

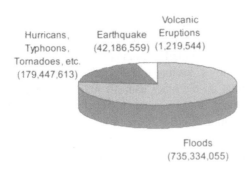

Figure 1.1. Consequences of the major disasters in the world (1966-1990): (a) number of human lives lost, and (b) number of people affected (source: Fattorelli et al., 1999)

1.1.2 A reminder of a few recent floods

Floods are news (Ward, 1978)! Almost every year major flood events hit some parts on earth. Dartmouth Flood Observatory (2003) has a chronological listing of major floods that occurred since 1985. A few such flood events and the damages they caused are presented here.

In the United States, the great flood of 1993 is one of the more remarkable floods during which 54,000 persons were evacuated, 50,000 homes were damaged and economic losses of US$15-20 billion have been estimated. The event also tested the limits of the nation's forecast and warning services as flood stages were exceeded at about 500 forecast points. Between 1965 and 1985, floods accounted for 63% of the federally declared disasters (337 out of 531), took 1,767 lives and caused US$5 billion worth of damage annually, on average. The extent of the danger of floods in the United States is revealed by the fact that there are 20,000 flood-prone communities: 3,000 of them receive site-specific flood forecasts from the National Weather Service (NWS) and 1,000 have local warning systems; the remaining communities receive country-wide warnings (Krzysztofowicz, 1995).

Europe has also been severely affected by floods for centuries. As examples of recent floods, in France 42 people died in 1992 during flash flooding in Vaison-la-Romaine; basin-wide floods caused widespread disruption and losses in the Rhine and Meuse basins in 1992, 1993 and 1995; and exceptional flooding struck the Po in 1994. In 1997 severe flooding occurred in several parts of Europe, both as localised flash floods and as basin-wide floods on major river systems causing loss of life, distress and disruption. The year started with flash flooding in Athens in mid-January and then in July exceptional rainfall in Czech Republic, Germany and Poland caused catastrophic flooding on the Oder River killing over 100 people and laying waste vast areas of countryside. Again, in early November, flash floods occurred, this time in Spain and Portugal with over 20 people losing their lives (Samuels, 1999). Remarkable floods of August 2002 severely hit central Europe, particularly many parts of Austria, Czech Republic and Germany with the losses of billions of euros.

Flooding is a major threat every year in many countries of Asia, including Bangladesh, China, Japan, India, Nepal, Thailand, and Vietnam. In Nepal alone, not to mention the social impact and property losses, 300 people die every year on average due to floods that occur together with landslides and debris flows. The exceptional monsoon rainfall in the year 2002 affected 46 of the 75 districts of Nepal including the capital city, 429 people died (Dartmouth Flood Observatory, 2003) and hundreds of houses were swept away.

1.2 Flood management options

The traditional approach of providing protection against flooding is through the construction of huge and costly structures by which floodplains are protected from more frequent inundation. Nowadays it has been widely realised that single objective flood "control" is inadequate due to a range of consequences (see, for example, Goodwin and Hardy, 1999; Grijsen et al., 1992; Rosenthal et al., 1998). It is relevant here to quote from the "Life on the Mississippi" (Chapter 28) by Mark Twain that describes the undefeated power of a natural river:

> "...that ten thousand River Commissions, with the mines of the world at their back, cannot tame that lawless stream, cannot curb it or confine it, cannot say it, Go here, or Go there, and make it obey; cannot save a shore which it has sentenced; cannot bar its path with an obstruction which it will not tear down, dance over, and laugh at."

The alternative is therefore to manage the flood rather than to fight against it. Flood management covers every aspect of flood mitigation and control measures (both structural and non-structural) at the pre-crisis and post-crisis stages as well as during the crisis. It also covers acquisition and application of information, reaction of people, political and social consequences and influences on decision making. Various means of flood hazard mitigation are described in Subsection 1.2.1, and a non-structural strategy called a Flood Forecasting, Warning and Response System (FFWRS) is presented in Subsection 1.2.2.

1.2.1 Alternative strategies of flood alleviation

A useful summary of various measures of flood alleviation that have been identified and applied in various flood mitigation schemes is listed in Figure 1.2. These measures are the result of research and development over many decades and even centuries (Penning-Rowsell and Peerbolte, 1994). They are categorised, conventionally, as either structural or non-structural. They can also be looked at as measures of water control, of land use control and of financial relief and loss reduction. Note that the same alternative may serve for two or more measures.

Figure 1.2 also substantiates the conclusion that floods are not just a physical phenomena: they are the result of our decisions to use the areas liable to flooding, and their impacts can be modified by the way in which we deploy the financial resources of our individuals and the state (Penning-Rowsell and Peerbolte, 1994). The subject of the present study looks at a component of a FFWRS, which is a non-structural and primarily a loss reduction measure.

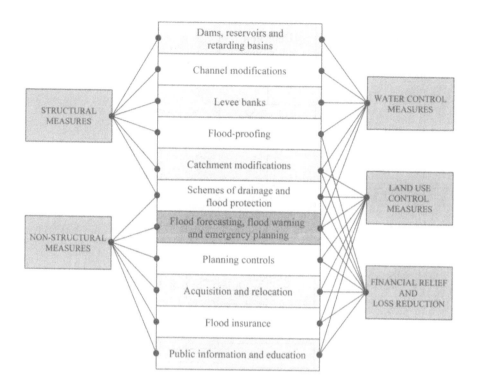

Figure 1.2. Alternative flood alleviation strategies (source: Penning-Rowsell and Peerbolte, 1994)

1.2.2 Flood forecasting, warning and response systems

The realisation that we must learn to live with floods emphasises the need for a reliable flood forecasting, warning and response system. The aim of this system is to reduce the loss of life and property by providing credible warnings to people at flood risk and to the authorities in charge of emergency flood protection and rescue operations. Regular river monitoring and river flood forecasting began with the advent of the telegraph, in 1854 in France, 1866 in Italy and 1871 in the USA (see Smith and Ward, 1998). A conceptual structure of a FFWRS (Parker et al., 1994) is presented in Figure 1.3. This extended structure is the result of the EUROflood (see Subsection 1.3.4) research project, which closely resembles the structure proposed by Krzysztofowicz and Davis (1983).

The information flow in the FFWRS is either intrinsic or extrinsic, and there are also important feedback loops. The intrinsic information flow is part of the FFWRS and incorporates 'official' warnings of the floods. The extrinsic information flow includes, on the other hand, 'unofficial' warnings used by the flood plain users (Parker et al., 1994).

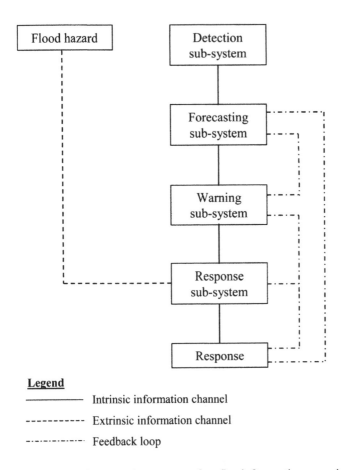

Legend

────────── Intrinsic information channel

---------- Extrinsic information channel

─·─·─·─·─ Feedback loop

Figure 1.3. Conceptual structure of a flood forecasting, warning and response system (FFWRS) (source: Parker et al., 1994)

The flood forecasting and flood warning sub-systems are two major components of the FFWRS, which are discussed further in this thesis. Flood forecasting is the process of estimating the conditions due to anticipated floods at a specific future time, or during a specific time interval. Flood warning, on the other hand, is the dissemination of forecasts to concerned authorities and to the general public with additional information attached to it, such as recommendations or orders for action. Flood warning is therefore a decision making process, which should take into account not only the forecasts but also the economical and social consequences. The benefits from warnings can be achieved only if they are timely, accurate, reliable and credible to the receivers, be they government agencies, businesses or private citizens.

1.3 Subject of the present research: uncertainty in flood forecasting and warning

Broadly, the subject of this study is about uncertainty in flood forecasting and warning systems. Management of uncertainty in the context of flood forecasting and warning implies a series of actions to be undertaken for the proper quantification of uncertainty in the forecasts and to make use of the uncertain forecasts in decision making (including flood warning and emergency planning). The actions should be focused to achieve a better consequence and normally involve (Maskey and Price, 2003a): (i) identification of the sources of uncertainty, (ii) reduction of uncertainty, (iii) quantification of uncertainty in inputs, (iv) propagation of uncertainty and the quantification of uncertainty in the output, and (v) use of the uncertain information in flood warning decision-making. A complete coverage of these steps for uncertainty management is beyond the scope of the present study. The focus here is on methods and methodologies for the propagation of various sources of uncertainty through a flood forecasting model for the quantification of uncertainty in flood forecasts. The frequent appearance of "warning" in this thesis is mainly to relate the usefulness of uncertainty quantification in practice.

The issue of uncertainty in flood forecasting and warning is further described in Subsection 1.3.1, and the potential benefits of uncertainty quantification are presented in Subsection 1.3.2. The aim and objectives of this research are presented in Subsection 1.3.3, and the two application examples that are part of this thesis are introduced in Subsection 1.3.4.

1.3.1 Uncertainty in flood forecasting and warning

In almost all circumstances, and at all times, we all find ourselves in a state of uncertainty (De Finetti, 1974). Uncertainty is therefore a common experience in everyday life and flooding is no exception. Indeed, all natural hazards are inherently uncertain in essence. Imperfect knowledge about the procedures and data generates uncertainty in the forecasts of floods. Various sources of uncertainty in flood forecasting are presented in Subsection 2.2.2. Errors in the forecast of the flood stage or of the time of arrival of flood conditions may lead (Smith and Ward, 1998) either

- to under-preparation and therefore to otherwise avoidable damage (if the forecast stage is too low and/or the forecast timing of inundation is late), or

- to over-preparation, unnecessary expense and anxiety, and to a subsequent loss of credibility (if the forecast stage is too high and/or the forecast timing of inundation is premature).

The fact that there exists uncertainty in flood forecasting raises issues about the reliability and credibility of flood warnings. A scientific solution to this problem is the implementation of uncertainty and risk-based flood warning systems. The risk-based design of civil engineering structures is increasingly common (see Tung and Mays, 1981; Van Gelder, 2000; Voortman, 2002; Vrijling 1993; Vrijling et al., 1998). Well formulated mathematical methods are available for risk-based flood warnings (Krzysztofowicz, 1993). The risk-based warnings are more rational, offer economical benefits and their needs are increasingly being realised (see Kelly and Krzysztofowicz, 1994; Krzysztofowicz, 1993; Krzysztofowicz et al., 1994). Proper evaluation of uncertainty is the basis for such systems.

Hiding the uncertainty associated with forecasts creates an illusion of certainty, the consequences of which may be very serious. An example of this is the flood event during the Spring of 1997 on the Red River in the Grand Forks, North Dakota, USA. The estimated 49 ft flood crest led city officials and residents to prepare as if this estimate were a perfect forecast (Krzysztofowicz, 2001a). The forecast was seriously underestimated as the actual flood crest of 54 ft overtopped the dikes, inundated 80% of the city and forced almost the total evacuation of its citizens. After the event, Eliot Glassheim, City Council Member in Grand Forks, North Dakota, observed (Glassheim, 1997):

"... the National Weather Service continued to predict that the river's crest at Grand Forks would be 49 ft...If someone had told us that these estimates were not an exact science, ...we may have been better prepared."

1.3.2 Benefits of uncertainty quantification in flood forecasting

Krzysztofowicz (2001a) presented various benefits of probabilistic forecasts in hydrology. In particular for flood forecasting, uncertainty levels attached to forecasts have the following benefits:

- They are scientifically more honest as they allow the forecaster to admit the uncertainty and to express the degree of certitude.

- They enable an authority to set risk-based criteria for flood warnings and emergency response.

- They provide information necessary for making rational decisions enabling the user to take risk explicitly into account.

- They offer the potential for additional economic benefits as a result of the forecasts to every rational decision maker and thereby to society as a whole.

- In the absence of rational decision making procedures (for example risk-based warnings), the information regarding uncertainty (for example the confidence intervals, the probability of exceedence of certain levels, etc.) helps decision-makers to use their own judgment more appropriately for decision making.

1.3.3 Aim and objectives of the present research

The aim of this research is to develop a framework and tools and techniques for uncertainty modelling in flood forecasting systems. The objectives of the research are:

- To identify the need for uncertainty assessment and to distinguish types and sources of uncertainty in flood forecasting;

- To review the theories of uncertainty representation and customary methods of uncertainty modelling;

- To develop a suitable framework and methodology, and propose improvements to existing method(s) for the better treatment of uncertainties, particularly the propagation of uncertainty through models, in flood forecasting;

- To explore the advantages and possibility of hybrid techniques of uncertainty modelling and probability-possibility (or fuzzy) transformations;

- To develop necessary computer codes to implement the developed methodology and to validate them by application to real-world problems.

1.3.4 Application examples

In recent years the European Commission (EC) has funded several projects involving studies of river basin and flood crisis management and mitigation including techniques for flood forecasting, warning and information dissemination. Some examples of recent EC funded projects are AFORISM (A Comprehensive Forecasting System for Flood Risk Mitigation and Control), 1991-1994; EUROflood, 1992-1994; RIBAMOD (River Basin Modelling, Management and Flood Mitigation), 1996-1998; TELEFLEUR (Telemetric Assisted Handling of Flood Emergencies in Urban Areas), 1998-2000; EFFS (European Flood Forecasting System), 2000-2003; and OSIRIS (Operational Solutions for the Management of Inundation Risks in the Information Society), 2000-2003.

The significance of uncertainty management in flood crisis was realised in the OSIRIS project, which resulted in the development of uncertainty assessment modules for the pilot flood forecasting systems of the project. Most of the present research was financed by this project. The application examples used in the present research are for two of the three pilot sites of this project namely, flood forecasting models for the Klodzko catchment in Poland and the Loire River in France.

1.4 Structure of the thesis

Chapter 2 reviews mathematical models used in flood forecasting and sheds some light on the issue of uncertainty assessment of the outputs from these models. This chapter also identifies uncertainty types and sources with respect to flood forecasting and reviews the commonly used uncertainty representation theories. Chapter 3 summaries methods of uncertainty analysis based on probability theory and fuzzy set theory and also reviews methods of probability-possibility transformations.

Chapter 4 presents the contribution of the present research in the field of uncertainty estimation in general and uncertainty in flood forecasting in particular. This involves the development of a methodology for the treatment of uncertainty due to time series inputs in flood forecasting (based on Monte Carlo technique and the fuzzy Extension Principle), an improvement to the existing first-order second moment method, development of qualitative uncertainty scales using best-case and worst-case scenarios and results of an investigation on hybrid techniques for modelling uncertainty and probability-possibility (or fuzzy) transformations. Chapters 5 and 6 present the application of the developed techniques in operational flood forecasting systems for the Klodzko catchment (Poland) and the Loire River (France), respectively. Chapter 7 is devoted to conclusions and recommendations. Appendix I includes definitions of fuzzy sets, defuzzification methods and fuzzy arithmetic. References are included at the end of the thesis.

Chapter 2

FLOOD FORECASTING MODELS AND UNCERTAINY REPRESENTATION

Summary of Chapter 2

This chapter describes different types of models used for flood forecasting as (i) physically-based, (ii) conceptual and empirical, and (iii) data-driven; and presents a discussion regarding the uncertainty assessment in these models. This chapter also defines uncertainty in rather general terms, distinguishing 'types of uncertainty' from 'sources of uncertainty'. Various types of uncertainty are discussed, such as *ambiguity* (one-to-many relationships), *randomness* and *vagueness* (or fuzziness), *inherent* and *epistemic* and *quantitative* and *qualitative*. Specific to flood forecasting, four sources of uncertainty are identified: input uncertainty, parameter uncertainty, model uncertainty and natural and operational uncertainty. Widely used theories of uncertainty representation are reviewed. The principles and introduction of probability theory, fuzzy set theory and possibility theory are presented.

2.1 Types of flood forecasting models

A conceptual structure of a typical flood forecasting, warning and response system (FFWRS) is presented in Chapter 1 (Fig. 1.3). The flood forecasting and flood warning sub-systems of the FFWRS are also defined in Subsection 1.2.2. A modelling system for real time flood forecasting consists of one or more of the following components: (i) a model of rainfall forecasting, (ii) a model of rainfall-runoff forecasting, and (iii) a model of flood routing and inundation (see, for example, Reed, 1984 and Yu and Tseng, 1996). Various aspects of flood modelling are presented by Price (2000). With the rapid development in hydrological and hydraulic modelling techniques and the necessary computer systems (both hardware and software), the trend these days is towards a so-called integrated real-time flood forecasting system, which consists of all three components of flood forecasting. It should however be noticed that in many parts of the world less sophisticated techniques of flood forecasting and ad-hoc warning systems are still in use.

The present study covers uncertainty analysis only in the rainfall-runoff and routing components of flood forecasting systems. Therefore, discussion regarding the models of rainfall forecasting is beyond the scope of this thesis. The model of rainfall-runoff forecasting also consists of a number of components depending on the level of detail used in the models. Here it is intended to cast some light on different types of available models for flood forecasting and a discussion vis-à-vis their uncertainty assessment. These models are commonly categorised on the basis of the extent to which they represent the physics of the processes involved. On this basis the flood forecasting models can be classified as

1. Physically-based
2. Conceptual and empirical
3. Data-driven or black box.

2.1.1 Physically-based models

The physically based models, as the name suggests, are based on the mathematical representation of all pertinent physical processes. The European Hydrological System or SHE (Abbott et al. 1986a & b) is an example of a systematic approach for the development of a physically based, fully distributed, catchment modelling system. SHE is physically based in the sense that the hydrological processes of water movement are modelled, either by finite difference representations of the partial differential equations of mass, momentum and energy conservation, or by empirical equations derived from independent experimental research (Abbott et al., 1986b). In fact, no models of a complex natural phenomenon like floods can be exclusively represented by mathematical equations and there exists some degree of approximation and empiricism. In this sense, even a sophisticated modelling system like SHE is not a complete representation of the physics of the natural phenomenon. However, the level of physical representation in SHE is such that it has been widely recognised as a standard physically based catchment modelling system.

Application of a distributed, physically-based model such as the SHE requires the provision of large amounts of parametric and input data, some of which may be time dependent (Abbott et al. 1986b). Such data are often unavailable. Problems may also arise due to an inadequate process representation in building a physically based model, for example, the representation of macropore flow as matrix flow. Moreover, such models handle large arrays of data and involve iterative solutions, which require considerable computing time and resources. Therefore, despite their theoretical advantages, engineers tend to rely more on conceptual and/or empirical models in many practical applications. There are however, some situations where physically based models can be very useful and should be preferred, for example, to assess the impact of changes in the physics of the catchment or for the assessment of extreme events (Guinot and Gourbesville, 2003).

2.1.2 Conceptual and empirical models

A conceptual model is generally composed of a number of interconnected conceptual elements. Conceptual models are not fully physically based, but their developments are inspired by the (limited) understanding of the physical processes. In other words these models endeavour to incorporate some aspects of the physical processes with some simplifications. There are many conceptual models with different levels of physical representation that have been widely used in hydrological forecasting. Fleming (1975) presented a thorough review of several conceptual models. Crawford and Linsley (1966) are credited for the development of the first major conceptual model by introducing the well known Stanford Watershed Model (SWM), which has undergone many modifications thereafter (Shamseldin, 2002). Some other widely used conceptual models include the Sacramento model (Burnash et al., 1973), the NAM model (Nielsen and Hansen, 1973), TOPMODEL (Beven and Kirkby, 1979; see also Beven et al., 1995), the TANK model (Sugawara et al., 1983), and so on.

The *empirical models*, on the other hand, relate input to the output assuming a very general relationship between the input and the output with little or no attempt to identify the physical processes involved. The empirical models are also referred to as *system-theoretical* models (Lettermann, 1991). Most of the *unit hydrograph* methods originally introduced by Sherman (1932) are widely used examples of empirical models. Although the derivation of this type of models is purely empirical the selection of the input-output parameters are, by and large, dictated by some physical understanding of the processes.

The conceptual and empirical models are normally lumped models in the sense that they use spatially averaged input data and parameters values. However, such models can be applied as semi-distributed models by dividing the catchments in to an appropriate number of sub-catchments.

2.1.3 Data-driven models

The basic idea of data-driven modelling techniques is to work with data only on the 'boundaries' of the domain where data are given, and to find a form of relationship(s) that best connects specific data sets (Price, 2002). The relationships can take a form that has little or nothing to do with the physical principles of the underlying processes. Traditionally the simplest of these models is the linear regression model. Nowadays there exists a host of nonlinear and sophisticated data-driven techniques, such as artificial neural networks (ANN), fuzzy rule-based systems (Bardossy and Duckstein, 1995), fuzzy regression (Bardossy et al., 1990), genetic programming (GP), support vector mechanics, etc (see, for example, Solomatine, 2002). Over the last decade some of these techniques have been used extensively, particularly in research, for water resources predictions, including rainfall-runoff modelling. Advantages of data-driven models are reported elsewhere (see, for example, Dawson and Wilby, 1998; Gautam and Holz, 2001; Dibike, 2002). The application of these relatively new techniques in operational forecasting systems is, however, still in an early stage.

2.1.4 Flood forecasting models vis-à-vis uncertainty analysis

Every model is, by definition, an approximation to reality (Price, 2002). A common feature of all these models (whether they be physically based or conceptual or data-driven) therefore, is that predictions from these models are far from being exact and they suffer from different degrees of uncertainty. Therefore, uncertainty quantification of the model result is essential, which is the subject of the present research. Uncertainty analysis is a well accepted procedure in the first two categories (physically based and conceptual and empirical) of model, although many operational flood forecasting systems may not have components for uncertainty analysis. Normally, increasing the complexity of the model, which generally means using more detailed mathematical representation of the physical processes, decreases uncertainty in the model output due to the lack of knowledge or incompleteness in conceptualising the real system (see Subsection 2.2.2 for sources of uncertainty). In this sense, model uncertainty should in principle decrease as we move from empirical to conceptual to physically based models. But model uncertainty is only a part of the total uncertainty. With the increasing complexity of the model, the number of inputs and parameters increases. If there exists large uncertainty in the inputs and if the parameters cannot be estimated with high precision, using complex models does not guarantee less uncertain model results. In physically based models uncertainty also arises due to the inadequate representation of the geometry and the schematisation of the processes. In fact, Guinot and Gourbesville (2003) showed that the uncertainty induced by the schematisation of the processes and of the geometry is much larger than that induced by the lack of knowledge of the parameters. Moreover, integration of uncertainty analysis with methods like Monte Carlo in models with a large number of parameter values and finer discretisation is very expensive in computer time. Beven (1989) presented more discussions on various issues of uncertainty assessment in physically based models.

Some of the data-driven techniques, such as the fuzzy rule-based systems and fuzzy regression, work with imprecise data and implicitly incorporate the uncertainty concept in modelling. These models however do not have the flexibility of using uncertainty methods based on other theories (e.g. the most popular probability theory), which is possible in the first two types of models. In the case of the ANN-based model (most popular so far of the data-driven techniques), model performance is, by and large, expressed in a form based on the difference between the observed and model predicted results using measures such as Root Mean Square Error (RMSE). Markus et al. (2003) used the *entropy measure* (Shannon, 1948) to quantify uncertainty in the results of ANN-based models. Methods for the application of more commonly used uncertainty assessment techniques, such as those based on probability and fuzzy set theories, are yet to be investigated for ANN-based models. Without the application of such uncertainty assessment techniques, the data-driven model forecasts cannot be used in uncertainty/risk-based decision makings that require all the uncertainties to be expressed explicitly in probabilistic terms. The uncertainty aspect may be one of the reasons for the fewer applications of the ANN-based models in operational systems.

2.2 Uncertainty types and sources

A unique definition of *uncertainty* is hard to find in the literature. It would not be a surprise if a book on this very subject starts and ends without it being defined (see also Zimmermann, 1997b). Because uncertainty exists virtually in all situations, its meaning is conceived as something so obvious. Perhaps what everybody understands by uncertainty is an antonym of certainty – anything that is not certain is uncertain. Zimmermann (1997a & b) also defined *uncertainty* with respect to *certainty*:

> *"Certainty implies that a person has quantitatively and qualitatively the appropriate information to describe, prescribe or predict deterministically and numerically a system, its behaviour or other phenomena."*

Situations that are not described by the above definition, shall be called *uncertainty*.

In the context of the present thesis, uncertainty may be thought of as a measure of the incompleteness of one's knowledge or information about an unknown quantity to be measured or a situation to be forecast.

Ross (1995) illustrated in a diagram (Fig. 2.1) that the presence of uncertainty in a typical problem is much larger compared to the presence of certainty. The collection of all information in this context is termed as the "information world" (represented by the circle in Fig. 2.1). If the uncertainty is considered in the content of this information, only a small portion of a typical problem might be regarded as certain or deterministic. The rest inevitably contains uncertainty that arises from the complexity of the system, from ignorance, from chance, from various classes of randomness, from imprecision, from lack of knowledge, from vagueness and so forth.

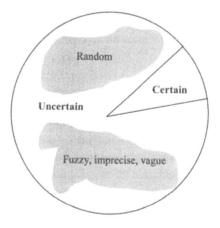

Figure 2.1. Certainty and uncertainty in the information world (source: Ross, 1995).

In this thesis, "types of uncertainty" are distinguished from "sources of uncertainty". The former is attached to the semantics and to measures with which the uncertainty in a body of information is characterised and quantified. The latter is used specifically in the context of modelling engineering systems to classify the sources that cause the uncertainty. Some of the classifications from the literature for the types of uncertainty are presented in Subsection 2.2.1. Sources of uncertainty in the context of flood forecasting are presented in Subsection 2.2.2.

2.2.1 Types of uncertainty

Various theorists in the field of uncertainty have proposed their own classifications for uncertainty. These all testify to the many faces of uncertainty and also to the difficulty of applying semantic classification to such an abstract concept (Hall, 1999). The more elaborate classifications endeavour to make subtle distinctions between different types of uncertain information. These are distinctions that may well be apparent in the minds of the authors but are not always well captured by the semantic labels attached to them. On the other hand, the less complex classifications have the attraction of simplicity (which is essential if key concepts are to be communicated to the practitioners) but may not do justice to some of the many faces of uncertainty. For simplicity, the high level classification by Klir and Folger (1988) of *ambiguity* (one-to-many relationships) and *vagueness* (or fuzziness) is preferable, but readers are advised to refer to the above citation if a more elaborate classification is intended. Their high level classification into ambiguity and vagueness coincides with that of Ayyub and Chao (1998). Ayyub (2001) defined *ambiguity* as the possibility of having multi-outcomes for processes or systems, and *vagueness* as noncrispness of belonging and nonbelonging of elements to a set or a notion of interest. The ambiguity type of uncertainty also includes the most appreciated *random* uncertainty (see Johnson and Ayyub, 1996). Owing to the dominant application of probability and fuzzy set theories in modelling uncertainty, uncertainty is also classified as *random* and *fuzzy* (e.g. Kaufmann and Gupta, 1991; Reddy and Haldar, 1992). In this thesis, in referring to uncertainty represented by probability and fuzzy set theories the random and fuzzy types of uncertainty are frequently used.

In the context of modelling engineering systems, the classification of uncertainty into *inherent* and *epistemic* (see, e.g. Van Gelder, 2000) is also very useful and appealing. The *inherent* uncertainty is the variability or randomness in space and time and the *epistemic* uncertainty is due to the lack of knowledge of fundamental phenomena. The latter type is further divided into statistical (parameter and distribution type) and model uncertainty. Uncertainty is also sometimes referred to as *quantitative* and *qualitative*. The former type refers to the uncertainty when it is represented using numerical values, typically by statistical inferences. Uncertainty if it is represented using linguistic variables is referred to as qualitative uncertainty.

2.2.2 Sources of uncertainty in flood forecasting

The sources of uncertainty refer to the causes that give rise in uncertainty in the forecasted state of a system that is modelled. In the context of flood forecasting, the sources of uncertainty are classified as (Maskey et al., 2003b):

1. Model uncertainty (due to assumptions in model equations, model building, and other forms of incompleteness in conceptualising the real system)

2. Input uncertainty (due to imprecise forecasts of and/or uncertainty in model inputs, such as future precipitation, evaporation, etc.)

3. Parameter uncertainty (due to imperfect assessment of model parameters)

4. Natural and operational uncertainty (due to unforeseen causes, e.g. glacier lake overflow, dam break, landslides, etc., malfunctioning of system components (hardware and software), erroneous and missing data, human errors and mistakes, etc.).

The above classification for the sources of uncertainty retains some features of the classification by Van Gelder (2000). Some other classifications that closely correspond to this classification are Kitanidis and Bras (1980): model uncertainty, input uncertainty, parameter uncertainty and initial state of the system; Melching (1995): natural randomness, data uncertainty, model parameters and model structure; and Krzysztofowicz (1999): input uncertainty, hydrological uncertainty and operational uncertainty.

The fourth category of the sources of uncertainty listed above is broad and is usually not addressed by uncertainty assessment in operational flood forecasting systems. The parameter uncertainty can reflect part of the uncertainty associated with the model structure (Cullen and Frey, 1999). Model uncertainty may also be assessed by comparing results from different models (Radwan et al., 2002). Such assessments of model uncertainty however, remain largely subjective.

The methods and methodologies developed and applied in this research are primarily focused on the treatment of input and parameter uncertainty. An exception is the framework of the expert judgement based qualitative method (Subsection 3.2.2), in which all identified sources of uncertainty can be represented qualitatively.

2.3 Theories of uncertainty representation

Historically, *probability theory* has been the primary tool for representing uncertainty in mathematical models (Ross, 1995). With the rapid development of computer technology and its use in mathematical modelling, the need of error representation in digital computation was increasingly realised in the 1950s and early 1960s, leading to the invention of *interval arithmetic* (Moore, 1962 & 1966). However, it was not until mid 1960s when Zadeh (1965) developed the *fuzzy set theory* that the representation of uncertainty by a non-probabilistic approach began to increase in pace rapidly. Further, Zadeh (1978) developed a broader framework for uncertainty representation called *possibility theory*, which is also known as a *fuzzy measure*. He interpreted a normal fuzzy membership function (Appendix I) as a possibility measure. Another broad theory of uncertainty representation was advanced by Shafer (1976) in the name of the *theory of evidence*. Shafer's theory has its origins in the work of Dempster (1969) on upper and lower probabilities. Therefore the theory of evidence is more commonly known as *Dempster-Shafer theory of evidence*. These theories are very rich in content, and therefore a detailed coverage of these cannot be presented within the scope of the present thesis. Some details on probability theory, fuzzy set theory and possibility theory (fuzzy measures) are presented in the following Subsections (2.3.1 to 2.3.3).

2.3.1 Probability theory

Of all the methods for handling uncertainty, probability theory has by far the longest tradition and is the best understood. This of course does not imply that it should be beyond criticism as a method of handling uncertainty. It does, however, mean that it is relatively well tested and well developed and can act as a standard against which other more recent approaches may be measured (Hall, 1999). There are two broad views on probability theory for representing uncertainty: *frequentist view* and *subjective or Bayesian view*.

Frequentist view of probability

The frequentist view of probability relates to the situation where an experiment can be repeated indefinitely under identical conditions, but the observed outcome is random. Empirical evidence suggests that the relative occurrence of any particular event, i.e. its relative frequency, converges to a limit as the number of repetitions of the experiment increases. This limit is what is called the probability of the event. The frequentist approach requires that we base statistical inferences on data that are collected preferably at random from a defined population. The mathematics of probability rests on three basic axioms. Let the probability of an event E be $P(E)$ and let X be the universal set that contains all possible elements of concern. The event here is also an element of the universal set X. The three axioms are the following:

Axiom 1: The probability of an event is always nonnegative, i.e.,

$$P(E) \geq 0 \tag{2.1}$$

Axiom 2: The probability of the universal set (or the sample space) is 1.0, i.e.,

$$P(X) = 1 \tag{2.2}$$

In terms of the individual element of the set, the Axiom 2 can also be expressed as:

$$\sum_{i=1}^{n} P(E_i) = 1 \qquad (2.3)$$

where n is the number of all elements of the set.

Axiom 3: If two events E_1 and E_2 are mutually exclusive, the probability of their union is equal to the summation of their probability, i.e.,

$$P(E_1 \cup E_2) = P(E_1) + P(E_2) \qquad (2.4)$$

It then follows in general for any events E_1 and E_2 that

$$P(E_1 \cup E_2) = P(E_1) + P(E_2) - P(E_1 E_2) \qquad (2.5)$$

Obviously, for mutually exclusive events, $P(E_1 E_2) = 0$ (Fig. 2.2), and Equation (2.5) reduces to Equation (2.4).

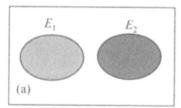 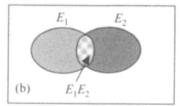

Figure 2.2. Venn diagram of (a) mutually exclusive events, and (b) intersection of two events.

Subjective or Bayesian view of probability

In the subjective view, probability is used as a belief. An event is a statement and the (subjective) probability of the event is a measure of the degree of belief that the subject has in the truth of the statement. The basic idea in the application of this approach is to assign a probability to any event on the basis of the current state of knowledge and to update it in the light of the new information. The conventional procedure for updating a prior probability in the light of new information is by using Baysian theorem.

Bayes' Theorem is named after Thomas Bayes, an 18th century mathematician (1702-1761) who derived a special case of this theorem. Bayes' theorem provides a rule for updating the belief in a hypothesis H (i.e. the probability of H) given additional evidence E and background information (context) I, as

$$P(H|E,I) = \frac{P(H|I)P(E|H,I)}{P(E|I)} \qquad (2.6)$$

The left hand-side term, $P(H|E,I)$, is called the *posterior* probability and it gives the probability of the hypothesis H after considering the effect of evidence E in the context I. The $P(H|I)$ term is just the *prior* probability of H given I alone; that is, the belief in H before the evidence E is considered. The term $P(E|H,I)$ is called the likelihood, which gives the probability of the evidence assuming the hypothesis H and background information I is true. The denominator of the right-hand term, $P(E|I)$, is the prior probability of the evidence that can be regarded as a normalising or scaling constant. This normalising constant is obtained by evaluating the exhaustive and exclusive set of evidence scenarios (Hall, 1999):

$$P(E \mid I) = \sum_{i=1}^{n} P(E \mid H_i, I) P(H_i \mid I) \tag{2.7}$$

The Bayesian theorem (Equation (2.6)) actually comes from a simple consequence of the definition of conditional probability. The conditional probability of two sets provides the dependency relationship between them. Given two sets A and B, the conditional probability states that

$$P(A|B) = \frac{P(A \cap B)}{P(B)} \tag{2.8}$$

A frequentist who has no data is paralysed (Cullen and Frey, 1999). However, there may be cases in which data are lacking in quantity or quality but for which an analyst has other information that can be used to construct a probabilistic representation of an input to a model. This is the kind of a situation the Baysian approach provides advantage over the frequentist approach. But this approach is not without shortcomings. In using a conventional Bayesian analysis the information concerning uncertain statistical parameters or the states of nature, no matter how vague, must necessarily be represented by conventional, exactly specified, probability distributions. Caselton and Luo, (1992) argues that such statistical precision may sometimes lead to the danger that inappropriately strong conclusions being drawn from the decision analysis.

Probability density function

In probability theory the uncertainty is represented by a Probability Density Function (PDF), $p_X(x)$, where X is the uncertain variable and x is its value. A cumulative form of the PDF is called a Cumulative Distribution Function (CDF), $P_X(x)$ (Fig. 2.3). The probability of X in the interval $(a, b]$ is given by

$$P(a < X \le b) = \int_a^b p_X(x)dx \tag{2.9}$$

It then follows

$$P_X(x) = P(X \leq x) = \int_{-\infty}^{+x} p_X(x)dx \qquad (2.10)$$

It also follows, if $P_X(x)$ has a first derivative, that

$$p_X(x) = \frac{dP_X(x)}{dx} \qquad (2.11)$$

It should be noted that any function representing a probability distribution of a random variable must necessarily satisfy the axioms of probability (Equations (2.1) – (2.3)). Therefore, a CDF possesses following properties (Ang and Tang, 1975):

1. $P_X(-\infty) = 0$; $P_X(+\infty) = 1.0$

2. $P_X(x) \geq 0$, and is nondecreasing with x, and

3. It is continuous with x.

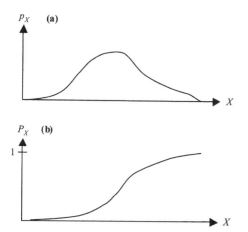

Figure 2.3. Example of (a) probability density function, and (b) cumulative distribution function.

2.3.2 Fuzzy set theory

In classical, or crisp, set theory any element x of the universal set X can be classified as being either an element of some sub-set A ($x \in A$) or an element of its complement ($x \in \overline{A}$, i.e. $x \notin A$). In other words, the transition for an element in the universe between membership and non-membership in a given set is abrupt and well-defined (hence called *crisp*) with membership either 1 (which certainly belongs to the set) or 0 (which certainly does not belong to the set). In many practical situations it is more logical to define the boundaries of sets vaguely so that an element can still be a member of a set with a degree of membership other than 1. This concept was first implemented by Zadeh (1965) with the introduction of *fuzzy set theory*. In fuzzy sets, therefore, the transition between the membership and non-membership can be gradual. This gradual transition of memberships is due to the fact that the boundaries of fuzzy sets are defined imprecisely and vaguely. This property of a fuzzy set makes the fuzzy set theory viable for the representation of uncertainty in a nonprobabilistic form.

Membership function

The essence of a fuzzy set, therefore, is the *membership* associated with the elements of the set. Membership is defined as the *degree of belief*, also called *belief level*, to which the element belongs to the set. The degree of belief or belief level or membership may take any value between and including 0 and 1: 0 meaning no membership and 1 meaning full membership.

In order to define a fuzzy set formally, let X be a universe set of x values (elements). Then $\tilde{A}$ is called a fuzzy (sub)set of X, if $\tilde{A}$ is a set of ordered pairs:

$$\tilde{A} = \left\{(x, \mu_{\tilde{A}}(x)); \ x \in X, \mu_{\tilde{A}}(x) \in [0,1]\right\} \tag{2.12}$$

where $\mu_{\tilde{A}}(x)$ is the grade of membership (or degree of belief) of x in $\tilde{A}$. The function $\mu_{\tilde{A}}(x)$ is called the *membership function* of $\tilde{A}$. A membership function (Fig. 2.4) maps every element of the universal set X to the interval [0, 1], i.e.

$$\mu_{\tilde{A}}(x): X \to [0,1] \tag{2.13}$$

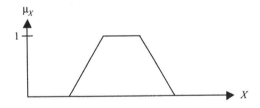

Figure 2.4. A normal (trapezoidal) membership function. (See Appendix I for the definition of a normal membership function.)

Basic operations on fuzzy sets

As in classical sets *union*, *intersection* and *complement* are the three basic and most often used operations in fuzzy sets. Given two fuzzy sets $\tilde{A}$ and $\tilde{B}$ the union, intersection and complement operations are represented as:

$$\text{Union}: \qquad \mu_{\tilde{A} \cup \tilde{B}}(x) = \mu_{\tilde{A}}(x) \vee \mu_{\tilde{B}}(x) \tag{2.14}$$

$$\text{Intersection}: \quad \mu_{\tilde{A} \cap \tilde{B}}(x) = \mu_{\tilde{A}}(x) \wedge \mu_{\tilde{B}}(x) \tag{2.15}$$

Complement : $\mu_{\tilde{\bar{A}}}(x) = 1 - \mu_{\tilde{A}}(x)$ (2.16)

The union and intersection of two fuzzy sets $\tilde{A}$ and $\tilde{B}$ (Fig. 2.5) and the complement of $\tilde{A}$ are illustrated in Figure 2.6(a-c).

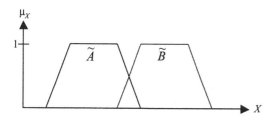

Figure 2.5. Fuzzy sets $\tilde{A}$ and $\tilde{B}$ represented by trapezoidal membership functions.

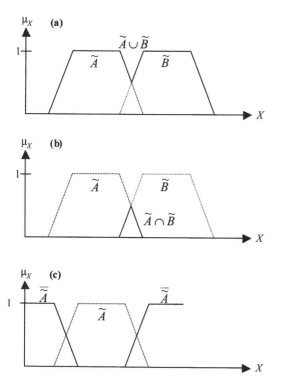

Figure 2.6. Operations between two fuzzy sets: (a) union, (b) intersection, and (c) complement.

A fuzzy set defined by a normal as well as convex fuzzy set is called a fuzzy number (see Appendix I for definitions). It is to be noted that the union and intersection operations may result in a non-normal and non-convex fuzzy set. Therefore, the operations on fuzzy numbers are normally carried out by fuzzy arithmetic. The rules of fuzzy arithmetic are based on the Extension Principle of fuzzy set theory. The Extension Principle for operations on fuzzy numbers is described in detail in Subsection 3.2.1, and the derivations for the basic fuzzy arithmetic are presented in Appendix I.

2.3.3 Possibility theory and fuzzy measures

In the late 1970s Zadeh advanced a theoretical framework for information and knowledge analysis, called *possibility theory*, emphasising the quantification of the semantic, context-dependent nature of symbols – that is, *meaning* rather than measure of information (Tsoukalas and Uhrig, 1997). The theory of possibility is analogous to, yet conceptually different from the theory of probability. Probability is fundamentally a measure of the frequency of occurrence of an event, while possibility is used to quantify the meaning of an event.

To understand the distinction between the probability and possibility of an event, it is worthwhile to observe an example similar to that given by Zadeh (1978). Suppose that we have a proposition "Hans ate X eggs for breakfast", where $X = \{1,2,3...\}$. The probability distribution, $p_X(x)$, and the possibility distribution, $r_X(x)$, associated with X may be represented as given in Table 1.

Table 1: An example of probability and possibility distributions of an event.

No of eggs (x)	1	2	3	4	5	6	7	8	9
$p_X(x)$	0.2	0.7	0.1	0	0	0	0	0	0
$r_X(x)$	1	1	0.8	0.6	0.3	0.2	0.1	0.0	0.0

The possibility distribution is interpreted as the *degree of ease* with which Hans eats x eggs. For example, the degree of ease with which Hans can eat two eggs is 1, whereas the degree of ease for eating seven eggs is only 0.1. The probability distribution might have been determined by observing Hans at breakfast for several days. As can be seen in Table 1, the probabilities in a probability distribution must sum to unity. The possibility distribution, on the other hand, is the *situation* or *context-dependent* and does not have to sum to unity. It is observed from the table that a high degree of possibility does not imply a high degree of probability. If, however, an event is not possible, it is also improbable. Thus, in a way, possibility is an upper bound for the probability. This weak connection between the two is known as *possibility/probability consistency principle* after Zadeh (1978).

Zadeh (1978) further postulated possibility as a *fuzzy measure*. A fuzzy measure is a function with a value between 0 and 1, indicating the *degree of evidence* or *belief* that a certain element x belongs to a set (Zadeh, 1978; Tsoukalas and Uhrig, 1997).

Therefore the membership function of a fuzzy event, say A, can be considered as a possibility distribution of the event. That is,

$$r_A(x) = \mu_A(x) \tag{2.17}$$

It should be noted however that only the normal membership function defines possibility measures (Bardossy and Duckstein, 1995). By virture of this connection between possibility distribution and the fuzzy membership function, the treatment of uncertainty using the fuzzy set theory is broadly known as a possibilistic approach (Langley, 2000).

Chapter 3

EXISTING MATHEMATICAL METHODS FOR UNCERTAINTY ASSESSMENT

Summary of Chapter 3

This chapter presents some existing mathematical methods for uncertainty assessments. In particular four methods are presented: Monte Carlo simulation, First-Order Second Moment method, fuzzy Extension Principle and the expert judgement-based qualitative method. The first two methods are based on probability theory and the latter two are based on fuzzy set theory. Brief principles of two other methods used in flood foresting problems are also presented. Primarily based on Bayesian theory, these two methods are the Bayesian Forecasting System and the Generalised Likelihood Uncertainty Estimation. There are growing concerns in the hybrid use of probability and possibility (or fuzzy set) theories to deal with the problems where the uncertainty comes from both randomness and vagueness. Various hybrid approaches are reported in the literature. Two of such approaches, namely, the concept of fuzzy probability and the concept of fuzzy-random variables are also presented. In dealing with the situation where the presence of both random and vague uncertainties is significant, the conversion from one representation to another is an important issue. Although several methods of conversions are reported, their acceptance in practical applications is yet to be seen. Two transformation methods, one based on a simple normalisation and another based on the principle of uncertainty invariance are also presented. At the end of this chapter, discussion is presented to illustrate some important distinctions and similarities of the uncertainty assessment methods presented at the beginning of the chapter.

3.1 Probability theory-based methods

The assessment of uncertainty in a model output requires the propagation of different sources of uncertainty through the model. The probability theory-based uncertainty assessment methods involve either the propagation of probability distributions or the moments of distributions (means and variances). Various methods exist both for the propagation of distributions and propagation of moments of distributions. These methods of uncertainty propagation can be broadly classified into three categories:

1. Analytical methods

2. Sampling methods

3. Approximation methods

Although various analytical solutions exist (e.g central limit theorem and transformation of variables) for the propagation of uncertainty through a model, they are limited to very simple problems and will not be detailed here. For problems with nonlinear functions of multiple variables, such solutions become complicated and are impracticable to use. Therefore, sampling methods, such as Monte Carlo (MC) and Latin hypercube simulations, or approximation methods, such as the First-Order Second Moment (FOSM) and Rosenblueth's point estimation (Rosenblueth, 1975) are preferred for practical applications. The sampling methods provide the estimation of the probability distribution of an output (propagation of distribution), while the approximate (or point estimate) methods provide only the moments of the distribution (propagation of moments). Comparative applications of different sampling methods and approximate or point estimate methods to a distributed rainfall-runoff model are reported by Yu et al. (2001). The MC simulation and FOSM are the broadly used methods for uncertainty assessment (Guinot, 1998) in a number of fields of engineering, including water resources. The present study also uses these two methods for the probabilistic assessment of uncertainty. Subsection 3.1.1 presents the principle of the Monte Carlo simulation and Subsection 3.1.2 outlines the principle of the FOSM method.

3.1.1 Sampling method: Monte Carlo simulation

The term Monte Carlo simulation was originally applied to methods of solving deterministic computational problems, for example, solving linear equations, differential equations and integrals, using statistical techniques. Nowadays, it is a term applied to any random sampling scheme employed by computers (Levine, 1971). It has also been used extensively and as a standard tool for probabilistic uncertainty assessment. In MC simulation, a model is run repeatedly to measure the system response of interest under various uncertain parameter sets generated from probability distributions of the parameters. The procedure for the application of MC simulation to assess a model output uncertainty is presented as following. Let the model under consideration consists of $X_1, ..., X_n$ as input random (uncertain) variables defined by their probability density functions (PDFs) and Y as output random variable.

1. Derive cumulative distribution functions (CDFs), $P_X(x_i)$, of all random variables ($i = 1, ..., n$) from their PDFs.

2. Generate n number of random numbers between 0 and 1. Let the random numbers be $u_1, ..., u_n$.

3. Determine a set of values of the random variables ($x_1, ..., x_n$) corresponding to the random numbers generated in Step 2 from their CDFs derived in Step 1. This is done by the method called *inverse transformation technique* or *inverse*

CDF method. In this method, the CDF of the random variable is equated to the generated random number, that is,

$$\left.\begin{array}{l} P_X(x_i) = u_i \\ x_i = P_X^{-1}(u_i) \end{array}\right\} \tag{3.1}$$

4. Evaluate the model for the set of input random variable values determined in Step 3, which results in a value of output random variable y_i.

5. Repeat Steps 2 to 4 as many times as required. This gives a set of outputs $y_1, ..., y_N$, where N is the number of repetitions or model runs.

6. From the set of outputs obtained in Step 5, derive the CDF and other statistical properties (e.g. mean and standard deviation) of Y.

The procedure described above is for uncorrelated random variables because the random numbers are generated independently for each of the variables (Step 2). Several methods are available for the treatment of correlation between random variables in to the framework of the MC simulation. The correlation coefficient is defined as the normalised covariance (see Subsection 3.1.2, Eq. (3.7)) between random variables. The correlation coefficient represents the degree of linear relationship between two variables. If two variables are independent they are also uncorrelated, but the converse is not necessarily true (Grimmett and Stirzaker, 1988). The detailed coverage of the treatment of correlated variables in MC simulations is beyond the scope of the present discussion, and the readers are referred to Cullen and Frey (1999), Haldar and Mahadevan (2000a) and Morgan and Henrion (1990) for a concise recipe on this topic.

Extensive applications of the Monte Carlo method for uncertainty analysis are reported in the literature. Some examples applied in uncertainty assessment of water resources systems, the list not exhaustive, are Chang et al. (1994), Gates et al. (1992), Krajewski and Smith (1995), Salas and Shin (1999), Song and Brown (1990), Van der Klis (2003), Warwick and Wilson (1990).

3.1.2 Approximation method: FOSM

The First-Order Second Moment method takes its name from the fact that it uses the first-order terms of the Taylor series expansion about the mean value of each input variable and requires up to the second moments of the uncertain variables. Owing to its simplicity, the FOSM method is one of the most widely used techniques in civil engineering applications for uncertainty assessment. This section describes the principle of the FOSM method.

Consider a function y that transforms a random variable X into a random variable Y, that is $Y = y(X)$. The exact mean or mathematical expectation $E(Y)$ and the variance, $Var(Y)$ of Y, are given by:

$$E(Y) = \int_{-\infty}^{+\infty} y(x) p_X(x) dx \qquad (3.2)$$

$$Var(Y) = \int_{-\infty}^{+\infty} [y(x) - E(Y)]^2 p_X(x) dx \qquad (3.3)$$

where p_X is the probability density function of X. To compute the mean (also called the first moment) and variance (or second moment) of Y using Equations (3.2) and (3.3), information on the p_X is needed. In many cases however, the available information is limited to the mean and variance of X. Furthermore, even if p_X is known, the computation of the integrals in Equations (3.2) and (3.3) may be time-consuming (Ang and Tang, 1975). Consequently, faster approximation methods are often preferred, that allow approximate values of the mean and variance to be computed. The FOSM method is one of these approximate methods.

For the sake of generality, consider a function of several random variables $X_1, \ldots, X_n$:

$$Y = y(X_1, \ldots, X_n) \qquad (3.4)$$

Expanding the function in a Taylor series about the mean values $\overline{X}_1, \ldots, \overline{X}_n$, yields the following expressions (see, e.g., Ang and Tang, 1975)

$$E(Y) = y(\overline{X}_1, \ldots, \overline{X}_n) \qquad (3.5)$$

$$Var(Y) = E\left[\left(\sum_{i=1}^{n}(X_i - \overline{X}_i)\frac{\partial y}{\partial X_i}\right)^2\right]$$

$$= \sum_{i=1}^{n}\left(\frac{\partial y}{\partial X_i}\right)^2 Var(X_i) + 2\sum_{i=1}^{n-1}\sum_{j=i+1}^{n}\left(\frac{\partial y}{\partial X_i}\right)\left(\frac{\partial y}{\partial X_j}\right)Cov(X_i, X_j) \qquad (3.6)$$

where $Cov(X_i, X_j)$ is the covariance between X_i and X_j, defined as

$$Cov(X_i, X_j) \equiv E[(X_i - \overline{X}_i)(X_j - \overline{X}_j)] \qquad (3.7)$$

All derivatives are evaluated at the mean values $\overline{X}_i$. The quantity $\partial y / \partial X_i$ is called the sensitivity of Y to the input variable X_i. The first term on the right-hand side of Equation (3.6) represents the contribution of the variances of the input variables to the total variance of the output. The second term denotes the influence of a possible correlation among the various possible pairs of input variables. If the input variables are statistically independent, i.e. $Cov(X_i, X_j) = 0$, this second term vanishes and the variance of Y becomes

$$Var(Y) = \sum_{i=1}^{n} \left(\frac{\partial y}{\partial X_i} \right)^2 Var(X_i) = \sum_{i=1}^{n} Var(Y)_i \qquad (3.8)$$

where $Var(Y)_i$ is the variance in Y due to the variance (uncertainty) in the input variable X_i. The standard FOSM method uses Equations (3.5) and (3.8) for the computation of the mean and variance, respectively, of the output variable when the input variables are statistically independent.

Although the method is simple and widely used, it suffers from some disadvantages (see Subsection 4.2.2). As a result of the present research an improvement has been applied to the conventional FOSM method. The result is an Improved FOSM method (Maskey and Guinot, 2002 & 2003), which is an original contribution of the present study and is described in detail in Section 4.2.

3.2 Fuzzy set theory-based methods

Uncertainty assessment is one of the various applications of the fuzzy set theory. Two methods of uncertainty assessment based on fuzzy set theory are discussed in this section: the Extension Principle-based method and the expert judgement-based qualitative method.

3.2.1 Fuzzy Extension Principle-based method

The fuzzy Extension Principle provides a mechanism for the mapping of the uncertain parameters (inputs) defined by their membership functions to the resulting uncertain output (dependent variable) in the form of a membership function. First developed by Zadeh (1975) and later elaborated by Yager (1986) this principle enables us to extend the domain of a function on fuzzy sets, which is the basis for the development of fuzzy arithmetic. In order to define this principle mathematically, consider a function of several uncertain variables $X_1, \ldots, X_n$ as:

$$Y = f(X_1, \ldots, X_n) \qquad (3.9)$$

Let fuzzy sets $\tilde{A}_1, \ldots, \tilde{A}_n$ be defined on $X_1, \ldots, X_n$ such that $x_1 \in X_1, \ldots, x_n \in X_n$. Equation (3.9) is identical to Equation (3.4), except that the uncertain variables in Equation (3.9) are fuzzy, instead of random in Equation (3.4).

The mapping of these input sets can be defined as a fuzzy set $\tilde{B}$:

$$\tilde{B} = f(\tilde{A}_1, ..., \tilde{A}_n) \tag{3.10}$$

where the membership function of the image $\tilde{B}$ is given by

$$\mu_{\tilde{B}}(y) = \begin{cases} \max\{\min[\mu_{\tilde{A}_1}(x_1), ..., \mu_{\tilde{A}_n}(x_n)]; y = f(x_1, ..., x_n)\} \\ \quad \text{if } \exists (x_i, ..., x_n) \in X_1 \times ... \times X_n \text{ such that } f(x_1, ..., x_n) = y \quad (3.11) \\ 0 \quad \text{otherwise} \end{cases}$$

The above equation is defined for a discrete-valued function f, if the function f is a continuous-valued then the max operator is replaced by the Sup (supremum) operator (the supremum is the least upper bound) (Ross, 1995).

Implementation of the Extension Principle by α-cut method

The direct application of the Extension Principle very often involves a computationally intensive procedure. Therefore, it is generally carried out in practice using the so-called α-cut method. By considering the fuzzy variable at a given α-cut level (see Appendix I), operations on fuzzy sets can be reduced to operations within the interval arithmetic (Dubois and Prade, 1980). The procedure of the implementation of the Extension Principle by the α-cut method can be explained as follows:

1. Select a value of $\alpha \in [0,1]$ of the membership function (the α-cut level).

2. For the given value of α find the lower bound $x_{i,LB}^{(\alpha)}$ and the upper bound $x_{i,UB}^{(\alpha)}$ of each fuzzy variable X_i for which $\mu_X(x_i) = \alpha$ (see Fig. 3.1).

3. Find the minimum and maximum values of $f(x_1, ..., x_n)$, considering all possible values of X_i located within the interval $[x_{i,LB}^{(\alpha)}, x_{i,UB}^{(\alpha)}]$. The minimum and maximum thus determined are the lower and upper bounds of the output Y for the given α-cut. That is,

$$\left. \begin{aligned} y_{min}^{(\alpha)} &= \min\{f(x_1, ..., x_n)\} \\ &= y_{LB}^{(\alpha)} \end{aligned} \right\}, \quad (x_1, ..., x_n) \in [x_{1,LB}^{(\alpha)}, x_{1,UB}^{(\alpha)}] \times ... \times [x_{n,LB}^{(\alpha)}, x_{n,UB}^{(\alpha)}]$$

$$\tag{3.12}$$

$$\left. \begin{aligned} y_{max}^{(\alpha)} &= \max\{f(x_1, ..., x_n)\} \\ &= y_{UB}^{(\alpha)} \end{aligned} \right\}, \quad (x_1, ..., x_n) \in [x_{1,LB}^{(\alpha)}, x_{1,UB}^{(\alpha)}] \times ... \times [x_{n,LB}^{(\alpha)}, x_{n,UB}^{(\alpha)}]$$

$$\tag{3.13}$$

The membership function of Y satisfies

$$\mu_Y(y_{LB}^{(\alpha)}) = \alpha \Big\}$$
$$\mu_Y(y_{UB}^{(\alpha)}) = \alpha \Big\}$$ (3.14)

4. Repeat Steps 1 to 3 for various values of α to construct the complete membership function of the output Y (see Fig. 3.2).

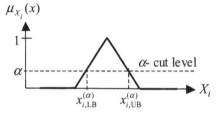

Figure 3.1. Membership function (arbitrary) of an input fuzzy variable X_i with its lower and upper bounds for a given α-cut level.

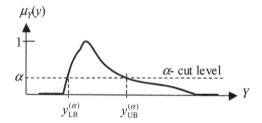

Figure 3.2. Membership function (arbitrary) of the output Y with the upper and lower bounds for a given α-cut level. The value of α is identical to that in Figure 3.1.

It can be seen from Equations (3.12) and (3.13) that finding the membership function of an output from the Extension Principle by the α-cut method amounts to finding the minimum and maximum of a function with constraints on the bounds in the input variables defined for the given α-cut. Various algorithms for finding minimum and maximum of a function are available, which range from linear, nonlinear to global. The suitability of the applications of these algorithms depends on the nature of the problem among other things. In the present study, global optimisation methods called genetic algorithms are used. More information on the algorithms is given in Section 5.3.

Recent applications of the Extension Principle in uncertainty analysis include groundwater flow problems (Schulz and Huwe, 1997 & 1999), chemical equilibrium problems (Schulz et al., 1999), environmental risk assessment (Guyonnet et al., 1999), and pipe networks analysis (Revelli and Ridolfi, 2002).

Extension Principle for monotonic function

If the given function f is strictly monotonic with respect to each input variable within the computation domain of the variables, the problem can be simplified as given in Equations (3.15) through (3.18).

$$y_{LB}^{(\alpha)} = f(x_{1,min}^{(\alpha)}, ..., x_{n,min}^{(\alpha)}) \qquad (3.15)$$

$$y_{UB}^{(\alpha)} = f(x_{1.max}^{(\alpha)}, ..., x_{n,max}^{(\alpha)}) \qquad (3.16)$$

where $x_{i,min}^{(\alpha)}, x_{i,max}^{(\alpha)}$ are defined as

$$x_{i,min}^{(\alpha)} = \begin{cases} x_{i,LB}^{(\alpha)} & \text{if } \dfrac{\partial f}{\partial x_i} > 0 \quad \forall (x_1,...,x_n) \in [x_{1,LB}^{(\alpha)}, x_{1,UB}^{(\alpha)}] \times ... \times [x_{n,LB}^{(\alpha)}, x_{n,UB}^{(\alpha)}] \\[3mm] x_{i,UB}^{(\alpha)} & \text{if } \dfrac{\partial f}{\partial x_i} < 0 \quad \forall (x_1,...,x_n) \in [x_{1,LB}^{(\alpha)}, x_{1,UB}^{(\alpha)}] \times ... \times [x_{n,LB}^{(\alpha)}, x_{n,UB}^{(\alpha)}] \end{cases}$$

$$(3.17)$$

$$x_{i,max}^{(\alpha)} = \begin{cases} x_{i,UB}^{(\alpha)} & \text{if } \dfrac{\partial f}{\partial x_i} > 0 \quad \forall (x_1,...,x_n) \in [x_{1,LB}^{(\alpha)}, x_{1,UB}^{(\alpha)}] \times ... \times [x_{n,LB}^{(\alpha)}, x_{n,UB}^{(\alpha)}] \\[3mm] x_{i,LB}^{(\alpha)} & \text{if } \dfrac{\partial f}{\partial x_i} < 0 \quad \forall (x_1,...,x_n) \in [x_{1,LB}^{(\alpha)}, x_{1,UB}^{(\alpha)}] \times ... \times [x_{n,LB}^{(\alpha)}, x_{n,UB}^{(\alpha)}] \end{cases}$$

$$(3.18)$$

This can be illustrated by taking a simple example of a uniform flow formula given by

$$Q = \frac{AR^{2/3} S^{1/2}}{N} \qquad (3.19)$$

where Q is the discharge, A is the area, R is the hydraulic radius, S is the slope of the water surface, and N is the Manning's roughness coefficient. Assuming A, R, S and N are uncertain parameters represented by their membership functions, the lower and upper bounds of the membership function for Q for a given α-cut are given by

$$Q_{LB}^{(\alpha)} = \frac{A_{LB}^{(\alpha)} (R_{LB}^{(\alpha)})^{2/3} (S_{LB}^{(\alpha)})^{1/2}}{N_{UB}^{(\alpha)}} \qquad (3.20)$$

$$Q_{UB}^{(\alpha)} = \frac{A_{UB}^{(\alpha)} (R_{UB}^{(\alpha)})^{2/3} (S_{UB}^{(\alpha)})^{1/2}}{N_{LB}^{(\alpha)}} \qquad (3.21)$$

3.2.2 Expert judgement-based qualitative method

Expert judgements play a significant role in the analysis of uncertainty. When there is lack of data to quantify uncertainties in various parameters, expert judgement may be used to make the best possible estimates of uncertainty. This is often the case in complex systems involving many uncertain parameters, the PDFs or moments of some of which cannot be determined explicitly by statistical observations due to insufficient data. Therefore, experts' judgements are used in one or other form to evaluate them. In general, experts are comfortable to make qualitative rather than quantitative judgement of such parameters. Fuzzy set theory, which uses linguistic (qualitative) variables to represent uncertainty, is very suitable to model uncertainty, particularly where experts' judgements are used extensively. The present method is fully based on expert judgements (using fuzzy variables) for the uncertainty in the input parameters, and it is qualitative also in the sense that it propagates the uncertainty by evaluating the model rather qualitatively. This method was originally reported by Sundararajan (1994 & 1998) to analyse the uncertainties in computed natural frequencies of nuclear power plant piping systems. Maskey (2001) and Maskey et al. (2002a) reported the method in the context of flood forecasting. The procedures involved in the application of this method are as follows:

Identification of sources of uncertainty and decomposition

As in any other methods of uncertainty analysis, the first step consists of identifying the sources of uncertainty in the output. These sources are then grouped into parameters and sub-parameters (decomposition of parameters). This allows the expert to assess the individual contributions of the decomposed parameters/sub-parameters rather than to assess the uncertainty of a combination of many parameters. Also, the quality of a parameter may be different for different events. This can be properly accounted for only if the parameter is broken down into sub-parameters. As an example, suppose for a particular event the rainfall for a basin was collected only from 4 gauge stations instead of usual 6 stations. The quality of the rainfall measurement in the 4-station case may be different from the 6-station case.

Assessment of the Quality and Importance of parameters/sub-parameters

An important concept used in this method is that the uncertainty contribution of each parameter is represented by two quantities: *Quality* and *Importance*. *Quality* refers to how good the knowledge we have for the estimation of this parameter, and *Importance* refers to how much this parameter contributes to the total uncertainty in the output. Evaluating the *Importance* of a parameter can be viewed as giving a weight to the parameter for its contribution to the total uncertainty in the output. If we compare *Quality* and *Importance* with the probability-based FOSM method, *Quality* is analogous to the variance of the parameter and *Importance* is analogous to the sensitivity of the output to the parameter.

Both *Quality* and *Importance* are evaluated qualitatively using a set of linguistic variables, each of which is represented by a membership function. For example, if five variables are used, the *Quality* may be evaluated as Very Good, Good, Acceptable, Bad and Very Bad. Equivalent linguistic variables for *Importance* may be Very Large, Large, Moderate, Small and Very Small. Membership functions are defined in arbitrary scale of 0 to 1. A typical set of linguistic variables (both for Quality and Importance) with their membership functions is shown in Figure 3.3.

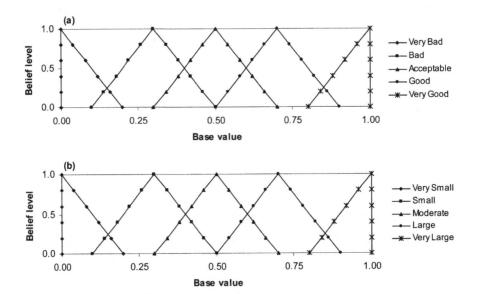

Figure 3.3. Membership functions of linguistic variables: (a) *Quality* and (b) *Importance*.

Derivation of parameters from sub-parameters

A parameter may be decomposed into several sub-parameters. In this case the importance is evaluated for the parameter, but the quality is assessed individually for the sub-parameters. The reason is that the quality of a parameter depends on the qualities of its sub-parameters, which may vary depending on the situations or events. It should be noted that the importance can also be derived from the importance of sub-parameters applying the similar procedure as for the quality. But care should be taken that the evaluated importance of the sub-parameter is for its contribution to the importance of its parameter not to the total uncertainty. The quality of the parameter can be derived by weighting the qualities of its sub-parameters represented by fuzzy numbers as given in Equation (3.22).

$$\Phi = (w_1.\phi_1(+)...(+)w_{n_s}.\phi_{n_s})$$
(3.22)

where Φ is the quality of a given parameter; ϕ_i ($i = 1,..., n_s$) is the quality of the sub-parameter i, n_s is the number of sub-parameters, and w_i is the crisp weight of the ith sub-parameter. The symbol (+) represents the addition between fuzzy numbers. The weights satisfy the following property:

$$\sum_{i=1}^{n_s} w_i = 1 \qquad (3.23)$$

Instead of supplying crisp weights (w_i) to sub-parameters, importance of each sub-parameter can be assessed using fuzzy variables. An alternative to assessing each sub-parameter quality is to use IF-THEN rules showing the relationships between the parameter and sub-parameter qualities set by the experts. Readers are referred to Sundararajan (1994 & 1998) for more details about this approach.

Combining Quality and Importance

The contribution of a given parameter to the total uncertainty is proportional to its *Importance*, Ψ. The higher the *Importance* of a parameter, the greater the uncertainty in the output. Conversely, the higher the *quality* of a parameter, the lower the resulting uncertainty. The contribution of a parameter to the total uncertainty is therefore proportional to the so-called "mirror image" Φ' of Φ. If the membership function of Φ is defined by couples $\{(x_1, s_1),..., (x_n, s_n)\}$, the membership of its mirror image is given by $\{(1-x_1, s_1),... (1-x_n, s_n)\}$, where $x_1,..., x_n$ are base values defined in the scale of 0 to 1 and $s_1,..., s_n$ are belief levels (i.e. $s_i = \mu(x_i)$).

The combined effect, U, of Φ and Ψ, is then given by

$$U_j = \Phi'_j(x)\Psi_j \qquad (3.24)$$

where the subscript j stands for the jth parameter, and the operator ($\times$) denotes fuzzy number multiplication.

Computing total uncertainty in the output

The total uncertainty in the output due to the uncertainty contributions of individual parameters is given by

$$U_O = \sum_{all\ j}^{fuzzy\ sum} \Phi'_j(x)\Psi_j \qquad (3.25)$$

where U_O is the total uncertainty in the output in the form of a membership function. When the evaluation is made by more than one expert, say m experts, then the fuzzy uncertainty from different experts given by Equation (3.25) are averaged (Eq. (3.26)) to get the final total uncertainty.

$$U_O = \frac{1}{m}(U_{O,1}(+)...(+)U_{O,m})$$ (3.26)

Interpretation of the output uncertainty

The output uncertainty given by Equation (3.25) or (3.26) is the fuzzy (or possibilistic) uncertainty represented by a membership function. The length of the support and the value obtained from defuzzification (Appendix I) give the idea of the amount of uncertainty represented by the fuzzy membership function, which is like using the first and second moments to express uncertainty in the probabilistic approach. The amount of uncertainty represented by the fuzzy membership function can also be looked at in terms of fuzziness and nonspecificity (Klir and Yuan, 1995). Maskey (2001) presented a comparison of uncertainty expressed in terms of the support, defuzzification by the centre-of-area and max-membership methods, nonspecificity and fuzziness.

As discussed earlier, the quality of a parameter is inversely proportional to the uncertainty contribution of the parameter. Similar is the case with the standard deviation (in probabilistic representation). Taking this as a basis Sundararajan (1994 & 1998) presented a comparison of the fuzzy presentation of uncertainty to standard deviation. This comparison however requires a known value of fuzzy uncertainty corresponding to a known value of a standard deviation, called a bench-marking value.

An alternative way is to express the uncertainty using linguistic variables, such as highly likely, likely, less likely etc., or expressing the uncertainty levels, such as small uncertainty, large uncertainty, etc. The present research proposed a Qualitative Uncertainty Scale to express the uncertainty levels, which is explained in Section 4.3.

3.3 Bayesian theory-based uncertainty assessment methods used in flood forecasting

Although various theories of uncertainty representation have shown good potential in other fields of engineering, only the methods based on probability theory have been used widely in flood forecasting problems. While the sampling method, MC simulation, and the approximate method, FOSM, have been widely used, the Bayesian forecasting system (Krzysztofowicz, 1999) and the generalised likelihood uncertainty estimation (Beven and Binley, 1992) can be considered as emergent methods in assessing uncertainty in flood forecasting. The latter two methods, primarily based on Bayesian theory, are briefly described here with recommendations to relevant literature for detail coverage. It should be noticed that these two methods also require sampling techniques like MC simulation.

3.3.1 Bayesian forecasting system (BFS)

The Bayesian forecasting system (BFS) by Krzysztofowicz (1999) is a recent development in the quantification of predictive uncertainty through a deterministic hydrological model. The BFS decomposes the total uncertainty about the variable to be forecasted (the "predictand") into input uncertainty and hydrological uncertainty. In BFS classification, the input uncertainty is associated with those inputs into the hydrological model which constitute the dominant sources of uncertainty and which therefore are treated as random and are forecasted probabilistically. The hydrological uncertainty is associated with all other sources of uncertainty such as model, parameter, estimation and measurement errors. The input uncertainty and the hydrological uncertainty are processed separately through the so-called input uncertainty processor and the hydrological uncertainty processor, respectively. These two uncertainties obtained from the two independent processors are then integrated by a so-called integrator. The BFS structure consisting of the input uncertainty processor, hydrological uncertainty processor and integrator is illustrated in Figure 3.4. The input uncertainty processor by the name of precipitation uncertainty processor (PUP) was further elaborated by Kelly and Krzysztofowicz (2000), and the hydrological uncertainty processor (HUP) by Krzysztofowicz and Kelly (2000). Similarly, the integrator was further exemplified by Krzysztofowicz (2001b).

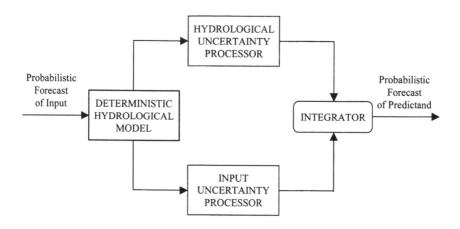

Figure 3.4. Structure of the Bayesian forecasting system (source: Krzysztofowicz, 1999).

3.3.2 Generalised likelihood uncertainty estimation (GLUE)

First introduced by Beven and Binley (1992), the generalised likelihood uncertainty estimation (GLUE) method reformulates the model calibration problem as the estimation of posterior probabilities of model responses (Romanowicz and Beven, 1998). The posterior probabilities of the parameter sets are determined using the Bayesian theory as (Box and Tiao, 1973)

$$f(\theta \,|\, \mathbf{z}) = \frac{f(\theta)L(\theta \,|\, \mathbf{z})}{f(\mathbf{z})} \qquad\qquad (3.27)$$

where $\mathbf{z}$ is the vector of observations, $f(\theta \,|\, \mathbf{z})$ is the posterior distribution (probability density) of the sets of parameter values θ given the data, $f(\theta)$ is the prior probability density of the parameter sets, $f(\mathbf{z})$ is a scaling factor and $L(\theta \,|\, \mathbf{z})$ is a likelihood function for the parameter set θ given the observation set $\mathbf{z}$ obtained from the forward modelling. This form assumes that the data $\mathbf{z}$ are fixed at their observed values while the parameters sets θ are treated as random variables (Wetherill, 1981). This allows the introduction of prior distributions for the parameters (Romanowicz and Beven, 1998).

As the new data become available, Equation (3.27) can be applied sequentially taking the existing posterior distributions (based on $n-1$ calibration periods) as the prior for the new data in the nth calibration period. That is

$$f(\theta \,|\, z_1,...,z_n) \propto f(\theta \,|\, z_1,...,z_{n-1})L(\theta \,|\, z_n) \qquad\qquad (3.28)$$

The predictive distribution of the model responses (water levels) is determined with the formulation of a model of prediction errors. Romanowicz et al. (1996) and Romanowicz and Beven (1998) assumed an additive error model ($\delta_t = y - y_t$, where y and y_t are realizations of simulated and observed water levels, respectively) and a Gaussian distribution of errors with first-order correlation. For details on the GLUE method, the readers are referred to Beven and Binley (1992) and Romanowicz et al. (1996). Further discussion on GLUE is found in Aronica et al. (1998), Beven and Freer (2001) and McIntyre et al. (2002).

3.4 Hybrid techniques in uncertainty modelling

Uncertainty in a given system or a parameter may constitute components of the major types of uncertainty: randomness and vagueness. The former is dealt with using the theory of probability and the latter using the theory of fuzzy sets. This suggests that uncertainty assessment using only one of the approaches may be incomplete. Various concepts, which we refer as hybrid techniques, have emerged whereby the combined use of both the probabilistic and fuzzy approaches are proposed. In-depth coverage of this topic is beyond the scope of this thesis. Kaufmann and Gupta (1991) presented more discussion on this issue. A brief description of the two of such concepts, which have been used in some applications, is presented here. These are the concept of *fuzzy probability* by Zadeh (1968 & 1984) and the concept of *fuzzy-random variable* by Ayyub and Chao (1998). The readers are referred to these references for more details of these concepts.

3.4.1 The concept of fuzzy probability

The concept of fuzzy probability, that is, the probability of a fuzzy event was first proposed by Zadeh (1968) and later elaborated by Zadeh (1984). The concept is also addressed by Dubois and Prade (1988), Ross (1995), Terano et al. (1992), Tsoukalas and Uhrig (1997) and Zimmermann (1991). The fuzzy probability concept considers the probability (uncertainty) of an event that is fuzzy. The probability of a fuzzy event, also known as fuzzy probability, is given by

$$P(\tilde{F}) = \int_X \mu_{\tilde{F}}(x) p_X(x) dx \qquad (3.29)$$

where $\tilde{F}$ is the fuzzy event on the universe X, $x \in X$, $P(\tilde{F})$ is the probability of the fuzzy event $\tilde{F}$, $\mu_{\tilde{F}}(x)$ is the membership function of the fuzzy event and $p_X(x)$ is a probability distribution.

In case of discrete events, with $X = \{x_1, ..., x_n\}$, the probability of the fuzzy event $\tilde{F}$ on X can be expressed as:

$$P(\tilde{F}) = \sum_{i=1}^{n} \mu_{\tilde{F}}(x_i) p(x = x_i) \qquad (3.30)$$

Since this is a summation of the probability, $p(x = x_i)$, multiplied by the degree to which x_i belong to the fuzzy event $\tilde{F}$, it can be seen that it is a concept that is easy to accept intuitively as the probability that the event $\tilde{F}$ will occur (Terano et al., 1992). The concept of fuzzy probability is distinct from that of second-order probability (i.e. a probability-value which is characterised by its probability distribution) and contains that of interval-valued probability as a special case (Zadeh, 1984).

3.4.2 The concept of fuzzy-random variable

The term *fuzzy-random variable* was used by Ayyub and Chao (1998) to define an uncertainty parameter that carries uncertainty from both fuzziness and randomness. The method proposed by them to treat a fuzzy-random variable considers the membership function as a weight function to the probability distribution function. If the uncertainty is only due to randomness, (that is, if the variable is purely random) both the mean and the standard deviation of its PDF are defined by crisp numbers. In contrast, for a fuzzy-random variable either mean or standard deviation or both can be considered as a fuzzy number defined by respective membership functions. The MF transferred to (equivalent) PDF is used as a weight function to the original PDF. To explain this procedure, let us consider a fuzzy-random variable x defined by its PDF, $p_X(x)$, with the mean $\bar{x}$ and standard deviation σ. Let us assume that the standard deviation is a crisp number but the mean is a fuzzy number defined by its membership function $\mu_{\bar{X}}(\bar{x})$. The MF is then transferred to a PDF $g_{\bar{X}}(\bar{x})$ such that

$$\int_{\bar{x}_{LB}}^{\bar{x}_{UB}} g_{\bar{X}}(\bar{x})d\bar{x} = 1 \qquad (3.31)$$

where $\bar{x}_{LB}$ and $\bar{x}_{UB}$ are the upper bound and lower bound of the MF of the fuzzy mean $\bar{x}$ at *support*. The transformation used here is the same as the transformation by simple normalisation as explained in Subsection 3.5.1. For convenience, the PDF is defined to be a conditional PDF, $p_{X|\bar{X}}(x|\bar{x})$. Thus, if $p_X(x)$ is normal the expression for the conditional PDF of $p_X(x)$ is given by

$$p_{X|\bar{X}}(x|\bar{x}) = \frac{1}{\sigma\sqrt{2\pi}}\exp\left[-\frac{1}{2}\left(\frac{x-\bar{x}}{\sigma}\right)^2\right] \qquad (3.32)$$

Then a joint PDF $f_{X,\bar{X}}(x,\bar{x})$ is defined by the multiplication of $p_{X|\bar{X}}(x|\bar{x})$ and $g_{\bar{X}}(\bar{x})$ as

$$f_{X,\bar{X}}(x,\bar{x}) = g_{\bar{X}}(\bar{x})p_{X|\bar{X}}(x|\bar{x}) \qquad (3.33)$$

Then the fuzzy-random PDF $f_X(x)$ of the fuzzy-random variable is defined as the marginal PDF of the joint PDF $f_{X,\bar{X}}(x,\bar{x})$, i.e.

$$f_X(x) = \int_{\bar{x}_{LB}}^{\bar{x}_{UB}} g_{\bar{X}}(\bar{x})p_{X|\bar{X}}(x|\bar{x})d\bar{x} \qquad (3.34)$$

3.5 Methods for probability–possibility transformation

Establishing relationship between probabilistic and fuzzy (or more widely possibilistic) representation of uncertainty has gained significant attention since Zadeh (1978). Such a relationship or the transformation between probabilistic and possibilistic uncertainty is a key issue in handling both types of uncertainty in a given system. Various situations that give rise to the existence of probabilistic and possibilistic uncertainties in an uncertain parameter and/or in a system of modelling or forecasting are discussed in Chapter 4 (Section 4.4). Different methods of uncertainty modelling based on these theories have their own advantages and relevancies. The challenge is therefore to utilise the advantages and relevancies of both theories opportunistically. To achieve this we need the capability to move from one theory to the other as appropriate. That is, we need to integrate the theories in the sense that the uncertainty represented in one theory can be converted by a justifiable transformation into an equivalent representation in the other theory (Klir, 1992).

Various transformation methods have been suggested in the literature. Klir and Wierman (1998) presented discussions on various transformation methods based upon various principles reported in the literature (see, e.g., Leung, 1982; Kaufmann and Gupta, 1991; Klir, 1992; Wonneberger, 1994), such as based on maximum entropy, insufficient reason, maximum specificity and different forms of the principle of possibility-probability consistency. Based on experience, an ad-hoc-type conversion was also proposed by Bardossy and Duckstein (1995). Klir (1992) proposed a method based upon uncertainty invariance principle arguing that the transformation must preserve the amount of uncertainty contained in the information. Zimmermann (1991) presented discussion on various aspects of probability-possibility relationships. Klir and Wierman (1998) presented good reviews of various transformation methods reported in the literature until 1997. Except from the normalisation requirements these methods substantially differ from one another.

Most of the conversion methods from probability to possibility or vice versa use the formula of the form:

$$r_X(x_i) = \lambda (p_X(x_i))^\beta, \quad \forall i = 1,...,n \tag{3.35}$$

where $r_X(x)$ and $p_X(x)$ are possibility and probability distributions, n is the number of discrete values of probability and possibility measures. The coefficients λ and β (λ, $\beta > 0$) are determined depending on the principles used for transformation. Two transformation methods are discussed here: a method based on simple *normalisation* and the more sophisticated method based on the *uncertainty invariance principle*.

3.5.1 Transformation by simple normalisation

In this method the power coefficient β (Equation (3.35)) is taken as unity. The normalisation is required to insure that

$$\sum_{x \in X} p_X(x) = 1 \tag{3.36}$$

$$\max_{x \in X} \mu_X(x) = \max_{x \in X} r_X(x) = 1 \tag{3.27}$$

Thus, the probability to possibility transformation and the possibility to probability transformations are given by Equations (3.38) and (3.39), respectively. The advantage of this method is its simplicity but it lacks a theoretical basis.

$$r_X(x_i) = \frac{p_X(x_i)}{\max\{p_X(x_1),...,p_X(x_n)\}} \tag{3.38}$$

$$p_X(x_i) = \frac{r_X(x_i)}{\sum\limits_{i=1}^{n} r_X(x_i)} \qquad (3.39)$$

3.5.2 Transformation by principle of uncertainty invariance

The principle of uncertainty invariance, which is based on measures of uncertainty in various theories, provides some basis for establishing meaningful relationships between uncertainty theories. In particular, the transformation from one theory, say T_1 to another, say T_2, with the principle of uncertainty invariance requires that (Klir and Yuan, 1995; Klir and Wierman, 1998):

1. The amount of uncertainty associated with a given piece of information or a given situation be preserved when we move from T_1 to T_2.

2. The degree of belief in T_1 be converted to its counterparts in T_2 by an appropriate scale.

The requirement 1 guarantees that no uncertainty is added or eliminated solely by changing the mathematical theory by which a particular phenomenon is formalised. Therefore the amount of uncertainty measured before and after the conversion based on the respective theories must remain equal. The requirement 2 guarantees that certain properties, which are considered essential in a given context (such as ordering or proportionality of relevant values) are preserved by the transformation.

The transformation under discussion here is from probability (defined by a PDF) to possibility (defined by a MF as a possibility distribution) and vice versa. The expressions for the uncertainty measures in the probability and possibility theories are given below. A detailed coverage on these measures is beyond the scope of this thesis. The readers are referred to, e.g. Klir and Yuan (1995) and Klir and Wierman (1998), Ayyub (2001) for more coverage on this topic.

Uncertainty measures in probability theory

When uncertainty is formalised in terms of probability theory, the only measure of uncertainty applicable is the *Shannon entropy measure* (Klir, 1992). The *Shannon entropy measure* given by Shannon (1948) measures uncertainty due to conflict that arises from probability mass function of a body of information (Ayyub, 2001). This uncertainty is also referred to as *strife* or *discord*. More specifically, it measures the average uncertainty (in bits) associated with the prediction of outcomes in a random experiment. The *Shannon entropy*, represented by $H(p)$, is given by

$$H(p) = -\sum_{i=1}^{n} p_X(x_i) \log_2(p_X(x_i)) \qquad (3.40)$$

As can be observed in the $H(p)$ function, its range is form 0 (for $p_X(x) = 1, x \in X$) to $\log_2|X|$ (for uniform probability distribution with $p_X(x) = 1/|X|$ for all $x \in X$).

For the continuous random variable, the Shannon entropy can be extended to

$$H(p) = -\int_a^b p_X(x) \log_2(p_X(x)) dx \tag{3.41}$$

where a and b are the two extremes (minimum and maximum) of x defined in the PDF, $p_X(x)$.

Uncertainty measures in possibility theory

The possibility theory here is defined as fuzzy measures considering the membership function as a possibility distribution. The uncertainty existence in possibility theory can be measured in terms of *nonspecificity* and *strife*. Indeed the possibilistic uncertainty is defined as the sum of nonspecificity and strife (Eq. (3.44)). The *nonspecificity* type of uncertainty arises from lack of specificity in choosing one alternative (the *true* one) from several alternatives. This type of uncertainty vanishes, and complete certainty is achieved, when only one alternative is present. Therefore, the amount of uncertainty associated with a set of alternatives may thus be measured by the amount of information needed to remove the uncertainty. The *nonspecificity* type uncertainty in possibility theory, $N(r)$ can be expressed as (Klir, 1992):

$$N(r) = \sum_{i=2}^n r_X(x_i) \log_2 \frac{i}{i-1} \tag{3.42}$$

The other type of uncertainty, *strife*, in possibility theory, which arises from conflicts among various sets of alternatives can be expressed as

$$S(r) = \sum_{i=2}^n [r_X(x_i) - r_X(x_{i+1})] \log_2 \frac{i}{\sum_{j=1}^i r_X(x_j)} \tag{3.43}$$

Therefore the total uncertainty $U_T(r)$ in possibility theory is given by

$$U_T(r) = N(r) + S(r)$$
$$= \sum_{i=2}^n r_X(x_i) \log_2 \frac{i}{i-1} + \sum_{i=2}^n [r_X(x_i) - r_X(x_{i+1})] \log_2 \frac{i}{\sum_{j=1}^i r_X(x_j)} \tag{3.44}$$

Equation (3.44) can be rewritten as

$$U_T(r) = \sum_{i=2}^{n} [r_X(x_i) - r_X(x_{i+1})] \log_2 \frac{i^2}{\sum_{j=1}^{i} r_X(x_j)} + \Theta \tag{3.45}$$

where

$$
\begin{aligned}
\Theta &= \sum_{i=2}^{n} r_X(x_i) \log_2 \frac{i}{i-1} - \sum_{i=2}^{n} [r_X(x_i) - r_X(x_{i+1})] \log_2 i \\
&= \sum_{i=2}^{n} r_X(x_i) \log_2 i - \sum_{i=2}^{n} r_X(x_i) \log_2(i-1) - \sum_{i=2}^{n} r_X(x_i) \log_2 i + \sum_{i=2}^{n} r_X(x_{i+1}) \log_2 i \\
&= -r_X(x_2) \log_2(1) - \sum_{i=3}^{n} r_X(x_i) \log_2(i-1) + \sum_{i=3}^{n+1} r_X(x_i) \log_2(i-1) \\
&= -\sum_{i=3}^{n} r_X(x_i) \log_2(i-1) + \sum_{i=3}^{n} r_X(x_i) \log_2(i-1) + r_X(x_{n+1}) \log_2(n) \\
&= r_X(x_{n+1}) \log_2 n
\end{aligned}
$$

$$\tag{3.46}$$

By convention, $r_X(x_{n+1}) = 0$, i.e. $\Theta = 0$. Hence substituting $\Theta = 0$ in Equation (3.45), the total uncertainty in possibility theory is given by

$$U_T(r) = \sum_{i=2}^{n} [r_X(x_i) - r_X(x_{i+1})] \log_2 \frac{i^2}{\sum_{j=1}^{i} r_X(x_j)} \tag{3.47}$$

Formula for transformation

For the transformation form probability to possibility or vice versa, ProU $\leftrightarrow$ PosU, the principle of uncertainty invariance requires that

$$H(p) = N(r) + S(r) = U_T(r) \tag{3.48}$$

And the transformations are given by

Probability to possibility transformation (ProU $\rightarrow$ PosU):

$$r_X(x_i) = \left(\frac{p_X(x_i)}{p_X(x_1)} \right)^{\beta} \tag{3.49}$$

Possibility to probability transformation (PosU $\rightarrow$ ProU)

$$p_X(x_i) = \frac{r_X(x_i)^{1/\beta}}{\sum_{k=1}^{n} r_X(x_k)^{1/\beta}} \tag{3.50}$$

For simplicity $x_1,...,x_n$ are ordered such that $p_X(x_i) \geq p_X(x_{i+1})$ and $r_X(x_i) \geq r_X(x_{i+1})$ for all $i = 1,...,n-1$. For both way transformations (Eqs. (3.49) and (3.50)), the value of β $(0 < \beta < 1)$ is determined from the equality of the Equation (3.48). The transformation procedure is also illustrated in Figure 3.5.

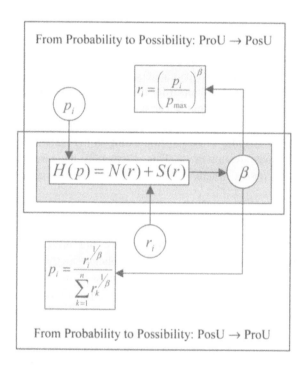

Figure 3.5. Probability-possibility transformations based on uncertainty invariance principle (source: Klir and Yuan, 1995). In the figure: $p_i = p_X(x_i)$, $p_{max} = \max\{p_X(x_1), ..., p_X(x_n)\}$, $r_i = r_X(x_i)$, and $r_k = r_X(x_k)$.

3.6 Discussion: probability– and fuzzy set theory–based methods

Probability theory and fuzzy set theory are the most widely used theories for uncertainty representation. Different methods for uncertainty assessment based on these two theories are presented in Sections 3.1 and 3.2. This section presents discussion on some important analogies and differences between these methods. In particular, the probability theory-based MC and FOSM methods are compared with the fuzzy set theory-based EP by the α-cut method and expert judgement-based qualitative method.

3.6.1   Monte Carlo simulation and Extension Principle

As mentioned previously, the Monte Carlo simulation is a widely used technique for uncertainty propagation through a model when the uncertainty in the parameters are characterised by probability distributions. On the other hand, the Extension Principle is used to propagate uncertainty when the parameters uncertainty are characterised by possibility distributions or membership functions. In the MC simulation the parameter values for each simulation is chosen randomly based on the probabilities of the parameters, which means that the scenarios that combine low probability parameter values have less chance of being randomly selected (Guyonnet et al., 1999). In the EP by the α-cut method, all possible combinations of parameters values are considered and the maximum and minimum model outputs obtained for the given intervals of the parameter values are directly reflected in the output uncertainty. Therefore, the EP-based method is by and large more conservative than the MC method. This suggests that the former is desirable when the extreme values corresponding to the parameter values at the tails of their distributions are important. This is generally the case in environmental and natural hazard context where human lives are often at stake (Guyonnet et al., 1999).

Another important distinction between the MC and the EP methods is the dependencies between the parameters. The MC method allows the effects of dependencies between the parameters to be accounted for via correlation coefficients that indicate the degree of correlation between parameters. In contrast, the current state of knowledge about the EP-based method does not allow the incorporation of the effect of correlation between the input parameters. In the EP by the α-cut method, we take the same α-cut level for all variables to determine the output uncertainty (interval) at the same α-cut level. This is not the same as correlation/uncorrelation in the MC method, and should be distinguished. In the simulation examples by Fishwick (1991), (fully) correlated and (fully) uncorrelated simulations were distinguished. In his correlated example, only two values (upper bound and lower bound) of each input are used to determine the interval in the output. This in fact is equivalent to the EP by α-cut for the case of a monotonic function (see Subsection 3.2.1). In the EP for non-monotonic function, all possible values of each parameter within the interval specified by the given α are considered. This case has been interpreted as (fully) uncorrelated by Fishwick (1991).

3.6.2   FOSM and expert judgement-based qualitative method

Although the FOSM method and the expert judgement-based qualitative method are based on two different theories (probability and possibility), they share some similarities. Observing Equations (3.8) and (3.25) we can see that the basic forms of these equations are same. Both are the summations of quantities resulting from the product of two quantities. The two quantities are the "square of the sensitivity" and the "variance" in the FOSM method and the "importance" and the "quality" (its mirror image) in the qualitative method. Therefore, in these two equations, the sensitivity of a parameter is equivalent to the importance of a parameter and the variance of a parameter is equivalent to the mirror image of the quality of the parameter (Fig. 3.6). The difference is only in the way the two multiplying quantities are represented and derived.

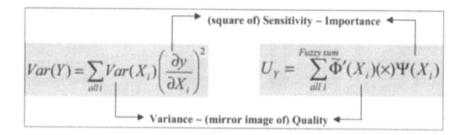

Figure 3.6. Comparison of equations for the FOSM method (probabilistic) and the expert judgement-based qualitative method. The symbols used in the figure are same as in Equations (3.8) and (3.25).

Chapter 4

CONTRIBUTION OF PRESENT RESEARCH TO UNCERTAINTY ASSESSMENT METHODS

Summary of Chapter 4

This chapter presents the original contributions of the present research to the development of uncertainty assessment methods. Firstly, an uncertainty assessment methodology using temporal disaggregation is presented. It aims to estimate the uncertainty in model output from the uncertainty in time series inputs. Three forms of uncertainty are considered: (i) uncertainty due to the temporal structure of the time series, (ii) uncertainty due to the spatial structure of the time series, and (iii) uncertainty due to the observed or forecasted magnitude of the input. Two algorithms are presented to implement the method in the frameworks of the fuzzy Extension Principle and the Monte Carlo simulation. The second contribution consists of an Improved First-Order Second Moment (IFOSM) method, which is derived using a second-order reconstruction of the model function. The improved method aims to correct the undesirable behaviour of the conventional FOSM method near extrema. The third contribution consists in developing qualitative uncertainty scales for the mapping of qualitative uncertainty estimated by the fuzzy set theory and expert judgement-based method presented in Chapter 3. Fourthly, the results of the investigation on hybrid approaches of modelling uncertainty and probability-possibility (fuzzy) transformations are presented. The differences and similarities in operation between random-random and fuzzy-fuzzy variables are particularly explored. It is showed that the addition and multiplication of two fuzzy variables by the Extension Principle using the α-cut method is similar to corresponding operations between two functionally dependent random variables for some specific conditions. It also provides an alternative method for the evaluation of the Extension Principle for a monotonic function without using the α-cut method. A transformation that is applicable in such cases is derived and illustrated by an example.

4.1 Uncertainty assessment methodology using temporal disaggregation

In water systems modelling, some of the forcing may be observed with a period larger than the typical reaction time of the catchment. A typical example is rainfall measured or forecasted on an hourly basis when the catchment response time is half an hour. In this case, using the (average) measured rainfall as the forcing functions directly as an input to the model may lead to an underestimate of the amplitude of the model response (because the variations in the signal are 'missed' and smoothed out into the averaged measurement). This is particularly true for peak values in the output that are often related to peak values in the inputs. In order to estimate better the peak values of the model output, the inputs must be reconstructed at a time scale smaller than the typical reaction time of the hydrological system under study. Failing to generate an input at a smaller time scale than that of the catchment response time may introduce error/uncertainty in the model outputs. If a model with multiple inputs is being used (e.g. a catchment model using records from several rainfall gages) the spatial distribution of the inputs is also a source of uncertainty.

Moreover, the average rainfall rate over a given subbasin is not necessarily equal to the point value measured at the corresponding rainfall gauge. Imprecision may also come from measuring devices. For the forecasted precipitation the uncertainty is unavoidable.

In the literature, the problem of time series reconstruction is more commonly referred to as temporal disaggregation. Some examples of disaggregation methods applied to precipitation data are Burian and Durrans (2002), Koutsoyiannis and Onof (2001), Margulis and Entekhabi (2001), Ormsbee (1989), Sivakumar et al. (2001) and Skaugen (2002). Similarly examples of application to stream flow time series are presented by Kumar et al. (2000) and Tarboton et al. (1998).

The purpose of temporal disaggregation, in the present discussion, is to assess the uncertainty in the output due to the "unknown" temporal distribution (of sufficiently higher frequency) of the input quantities. The methodology presented here aims at estimating the uncertainty in the output due to the uncertainty in time series inputs. It takes into account the three forms of uncertainty: (i) uncertainty due to the temporal structure of the time series, (ii) uncertainty due to the spatial structure of the time series, and (iii) uncertainty due to the observed or forecasted magnitude of the input. Two algorithms are presented for the application of the methodology: one based on the fuzzy Extension Principle and the other based on Monte Carlo simulation. The methodology is an original contribution of the present research (Maskey et al., 2003a & b; Maskey and Price, 2003b).

The principle of disaggregation is addressed in Subsection 4.1.1. The spatial variation of the temporal structure is addressed in Subsection 4.1.2. The uncertainty in the measured or forecasted precipitation is detailed in Subsection 4.1.3. Subsections 4.1.4 and 4.1.5 describe the synthesis of these three issues in the form of algorithms for fuzzy and probabilistic approaches respectively. Methods for the generation of disaggregation coefficients are presented in Subsection 4.1.6.

4.1.1 Principle of disaggregation

The basic principle of this method is to divide the given temporal period into a fixed number of subperiods and to randomly disaggregate the known accumulated sum of the time series input variable for the period into a number of subperiods, which aggregate to the given accumulated sum. The disaggregated values distributed over the subperiods are then used in the rainfall-runoff model as inputs.

Let W be the total quantity of a time series variable for a time interval T, called a period hereafter. If a disaggregated signal for a subperiod j is denoted by w_j then,

$$W = \sum_{j=1}^{n} w_j \tag{4.1}$$

where j is the index for the subperiods ($j = 1, ..., n$) and n is the number of subperiods.

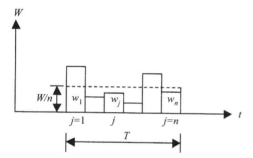

Figure 4.1. Average and disaggregated time series input.

For the sake of simplicity coefficients b_j are introduced, such that

$$w_j = W b_j \tag{4.2}$$

where

$$\left. \begin{array}{l} 0 \le b_j \le 1 \qquad \forall j \\ \displaystyle\sum_{j=1}^{n} b_j = 1 \end{array} \right\} \tag{4.3}$$

The coefficients b_j are generated randomly subject to the constraints (4.3). The Equations (4.2) and (4.3) allow the quantity w_j to take any value from 0 to W. An unbiased random generation of b_j provides all possible combinations of disaggregated signals, also referred to as the temporal patterns of the variable. The various disaggregated signals (temporal patterns) are then used as inputs to the model function

(Eq. (4.4)) to assess the uncertainty in the output, y, using either a probabilistic or a fuzzy approach.

$$y = f(w_j; j = 1,...,n)$$
$$= f(w_1,..., w_n)$$

(4.4)

A special case of the disaggregated pattern, which is referred to as the principal pattern by Maskey et al. (2003a), occurs when the coefficients take the value 1 for one of the subperiods, say j^*, and zero for the rest, i.e.

$$\left. \begin{array}{l} b_{j^*} = 1 \\ b_j = 0 \ \forall (j <> j^*) \end{array} \right\}$$

(4.5)

Consequently, from Equations (4.2) and (4.5),

$$\left. \begin{array}{l} w_{j^*} = W \\ w_j = 0 \ \forall (j <> j^*) \end{array} \right\}$$

(4.6)

It should be noticed, however, that this principal pattern of temporal distribution does not necessarily correspond to the extreme (upper bound or lower bound) values of the output.

Note that this methodology is based on the implicit assumption that the length of a subperiod is larger than the correlation time of the input signal. If it is not the case, the coefficients b_j should not be generated independently from each other. The length of the subperiod, however should not be too large, in which case the resulting signals will be close to the average signals over the period. Note that methods exist, that allow signals to be generated taking into account the signal correlation (Rodriguez-Iturbe and Eagleson, 1987; Mantoglou and Wilson, 1982). In the absence of any information about the correlogram of the input signal, such methods cannot be used and the coefficients b_j are to be generated independently from each other. However, due to the requirement of Equation (4.3), these coefficients cannot be fully independent. Various methods for the generation of the disaggregation coefficients are discussed in Subsection 4.1.6.

Also note that the number of subperiods has a direct influence on the maximum possible intensity of the input (e.g. maximum rainfall intensity). This is because the maximum possible intensity is obtained when the given value W for the period T is concentrated over one subperiod, with an intensity nW/T. Too many subperiods not only increases the computational requirements (see Subsection 4.1.4 and 4.1.5 for the algorithms) but may also lead to physically unfeasible results. Therefore, care must be taken in selecting the appropriate number of subperiods. A more rational solution to this problem may be achieved by constraining the disaggregation using more catchment specific information on the rainfall intensity and pattern. For example, relationships between the maximum intensity and the duration of the rainfall can be derived. Analysis of historical rainfall patterns may lead to the identification of several most probable rainfall patterns for various intensities for the region.

4.1.2 Spatial variations of temporal distribution

If we are to consider more than one point in space, the temporal distributions at different points in space can be different. For example in case of a catchment, the temporal distribution of precipitation over subbasins can be different. Denoting by m the number of subbasins, Equations (4.1) – (4.4) can be rewritten as

$$W_i = \sum_{j=1}^{n} w_{i,j}; \quad i = 1,...,m \tag{4.7}$$

$$w_{i,j} = W_i b_{i,j} \tag{4.8}$$

$$\left. \begin{array}{ll} 0 \le b_{i,j} \le 1 & \forall (i,j) \\ \sum_{j=1}^{n} b_{i,j} = 1 & \forall i \end{array} \right\} \tag{4.9}$$

$$\begin{aligned} y &= f(w_{i,j}; i = 1,...,m; j = 1,...,n) \\ &= f[(w_{1,1},...,w_{1,n}),...,(w_{m,1},...,w_{m,n})] \end{aligned} \tag{4.10}$$

where W_i is the accumulated value of the input for subbasin i and $w_{i,j}$ are the disaggregated signals for subbasin i and subperiod j. Whereas varying the coefficient $b_{i,j}$ over subperiods allows the possibility of different temporal distributions, varying the coefficient over subbasins allows for different spatial distributions. Thus by generating different values of $b_{i,j}$, as many temporal and spatial patterns as are needed can be generated.

4.1.3 Uncertainty in the input quantity

The principles described so far deal with the uncertainty due to the unknown temporal distribution of a variable over the subperiods and subbasins. There can be uncertainty also in the quantity of the input (accumulated sum) measured or forecasted over the period. This uncertainty is normally represented by a probability density function (PDF). Such a PDF is either derived from the available data or assumed in the absence of a sufficient number of data values. In case of the fuzzy set theory-based approach, the uncertainty is represented by a membership function (MF). Arbitrary representations of the uncertainty in the magnitude of the input (accumulated sum) by a PDF and a MF are shown in Figure 4.2.

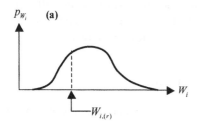

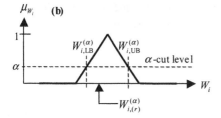

Figure 4.2. Representation of uncertainty (arbitrary) in time series inputs (W_i) accumulated over a period T by (a) a PDF, and (b) a MF.

4.1.4 Algorithms for the fuzzy approach

This subsection presents the implementation of the disaggregation principle in the form of an algorithm for a fuzzy set theory-based approach. The fuzzy Extension Principle is used (see Subsection 3.2.1) to implement this approach. The proposed methodology is independent of the structure of the forecasting model and can be used with any deterministic model. An application of this methodology is presented in Chapter 5 in the context of a rainfall-runoff type of flood forecasting model with the precipitation as the time series input. Assuming that the uncertainties in the accumulated sums W_i are represented by fuzzy membership functions (Fig. 4.2b), the algorithm is the following:

1. Select an α-cut ($\alpha \in [0,1]$) for the membership functions of all the model inputs. Take the same α for all membership functions.

2. For a given α-cut, determine for all $i = 1,\ldots, m$ the lower bound, $W_{i,\text{LB}}^{(\alpha)}$, and the upper bound, $W_{i,\text{UB}}^{(\alpha)}$.

3. Generate randomly a value $W_{i,(r)}^{(\alpha)}$ between $W_{i,\text{LB}}^{(\alpha)}$ and $W_{i,\text{UB}}^{(\alpha)}$ for all $i = 1,\ldots, m$ (see Fig. 4.2b). This value corresponds to the accumulated sum for the period T, i.e. W_i in Equation (4.1).

4. Generate coefficients $b_{i,j,(s)}$ between 0 and 1 ($i = 1,\ldots, m, j = 1,\ldots, n$) using one of the methods described in Subsection 4.1.6.

5. Use the values $W_{i,(r)}^{(\alpha)}$ generated in Step 3 and the coefficients $b_{i,j,(s)}$ generated in Step 4 as inputs in Equation (4.8) to redistribute the quantity over the subperiods, i.e.

$$w_{i,j,(r),(s)}^{(\alpha)} = W_{i,(r)}^{(\alpha)} b_{i,j,(s)} \tag{4.11}$$

6. Use the $w_{i,j,(r),(s)}^{(\alpha)}$ ($i = 1, ..., m; j = 1,..., n$) as inputs to the model (Equation (4.10)). This produces an output $y_{(r),(s)}^{(\alpha)}$.

7. Repeat Steps 4 to 6 s times. This means that for every value of the input generated in Step 3, s disaggregated sets of inputs are generated. This produces s outputs.

8. Repeat Steps 3 to 7 r times. This produces $s \times r$ outputs. From the $s \times r$ outputs determine the minimum and maximum of the outputs. The minimum and maximum values correspond to the lower bound and upper bound of the outputs, respectively, i.e.

$$\left.\begin{aligned} y_{LB}^{(\alpha)} &= \min_{r,s}(y_{(r),(s)}^{(\alpha)}) \\ y_{UB}^{(\alpha)} &= \max_{r,s}(y_{(r),(s)}^{(\alpha)}) \end{aligned}\right\} \tag{4.12}$$

The values of s and r are primarily governed by the size of the search space for the determination of minimum and maximum values of the outputs. The search space in this example is the function of m and n. Obviously, more subbasins and subperiods means more possible combinations of input sets. Different types of algorithms for the determination of the minimum and maximum values may deal differently for the values of s and r, such as using stopping criteria. In this study, genetic algorithms are used for the determination of the optima and are presented in Section 5.3.

9. Repeat Steps 2 to 8 for as many α-cuts as are needed to produce a complete MF for the output.

4.1.5 Algorithm for the Monte Carlo approach

For the probabilistic assessment of uncertainty, the methodology can be used in the framework of the Monte Carlo simulation (see Subsection 3.1.1). Assuming that the uncertainties in the accumulated sums W_i are represented by PDFs (Fig. 4.2a), the algorithm used for the methodology is the following:

1. Generate randomly a value $W_{i,(r)}$ based on its PDF (see Subsection 3.1.1) for all $i = 1, ..., m$. This value corresponds to the accumulated sum over the period T, i.e. W_i in Equation (4.1).

2. Generate coefficients $b_{i,j,(s)}$ between 0 and 1 ($i = 1,\ldots, m, j = 1,\ldots, n$) using one of the methods described in Subsection 4.1.6.

3. Use the value $W_{i,(r)}$ generated in step 1 and the coefficients $b_{i,j,(s)}$ generated in Step 2 in Equation (4.8) to redistribute the quantity over the subperiods, i.e.

$$w_{i,j,(r),(s)} = W_{i,(r)} b_{i,j,(s)} \tag{4.13}$$

4. Use the $w_{i,j,(r),(s)}$ ($i = 1, \ldots, m; j = 1,\ldots, n$) as inputs to the model (Equation (4.10)). This produces an output $y_{(r),(s)}$.

5. Repeat Steps 2 to 4 s times. This produces s outputs for a set of $W_{i,(r)}$ generated in Step 1.

6. Repeat Steps 1 to 5 r times. This produces $s \times r$ outputs. Determine the frequency distribution and other statistical properties of the output from the $r \times s$ sets of inputs-outputs. The values of r and s depend on the values of n and m, the degree of non-linearity and the level of details required for the output distribution.

4.1.6 Generation of the pattern (disaggregation) coefficients

In this methodology, the generation of the disaggregation coefficient (b_j) needs further elaboration. When the number of subperiods (n) for disaggregation is more than 1, and since the summation of the coefficients must be 1, b_j cannot be generated as fully independent random variables. Therefore, the way the coefficients are generated may influence the results. The following methods are identified:

1. By rejection
2. By symmetry
3. By normalization

The first method, by rejection, implies that the first $n - 1$ coefficients are randomly generated independently. The sum of the generated coefficients is then checked to see if it exceeds 1. If this is true, the whole set is rejected and a new set is generated until the sum remains equal to or less than 1. The nth coefficient is then calculated using Equation (4.14). The implementation of this method is presented in Procedure (1).

$$b_n = 1 - \sum_{k=1}^{n-1} b_k \tag{4.14}$$

```
Procedure (1): PatternCoef by Rejection
begin
   sum ← A (A>1);
   while (sum < 1) do
   begin
      generate b₁, ..., bₙ₋₁ randomly;
      sum ← b₁₊ ...+bₙ₋₁;
   end
   bₙ ← 1-(b₁₊ ...+bₙ₋₁);
end
```

Taking an example of three subperiods ($n = 3$), the resulting frequency diagrams of b_j are shown in Figure 4.3.

The second approach, by symmetry, is presented for the case $n = 3$. In this case, the first two coefficients are generated randomly and independently. A check is applied to see if the sum of the two exceeds 1. If this is true, instead of rejecting the set as in the "by rejection" method, the two coefficients are replaced by their respective complements from 1. This condition guarantees that the sum of the two remains less than 1. The 3rd coefficient, which is the nth coefficient, is then determined using Equation (4.14). Procedure (2) presents the implementation of this method.

```
Procedure (2): PatternCoef by Symmetry
begin
   generate b₁,b₂ randomly;
   sum ← b₁+b₂;
   If sum > 1 then
      b₁ ← 1-b₁;
      b₂ ← 1-b₂;
   endif
   b₃ ← 1-(b1+b2);
end
```

The resulting distribution of the 3 coefficients by this method is shown in Figure 4.4. It is interesting to see that the distributions from "by rejection" and "by symmetry" are same for each coefficient. The advantage of the second method over the first is that it is quicker as it needs only a single generation of the coefficients for each run, whereas, in the first method many sets of coefficients may need to be rejected before getting one that fits the condition.

The third method, by normalisation, implies that all the coefficients are generated independently and then normalised to obtain a sum of 1. The implementation of this method is presented in Procedure (3). In this method each coefficient needs to be generated only once for each run.

Procedure (3): PatternCoef by Normalisation
begin
    generate b_1, ..., b_n randomly;
    sum ← b_1+...+b_n;
    b_1 ← b_1/sum;
    .
    .
    .
    b_n ← b_n/sum;
end

The resulting distributions (Fig. 4.5) are markedly different from the first two methods. The normalisation method results in a distribution that is close to normal for each coefficient.

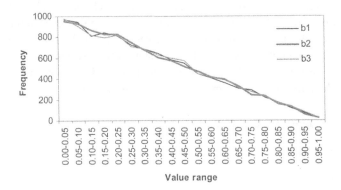

Figure 4.3. Relative frequency of b_j generated by rejection.

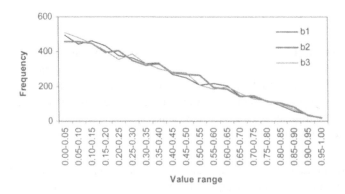

Figure 4.4. Relative frequency of b_j generated by symmetry.

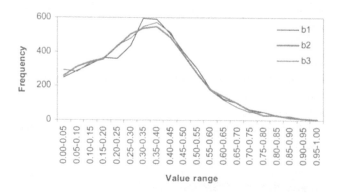

Figure 4.5. Relative frequency of b_j generated by normalisation.

There is one thing common to the resulting distribution for each method: as b_j (for all j) tends to 1 the probability of it being generated tends to 0. In the first two methods the probability is decreases linearly as the value of the coefficient increases. The question then is "what are the implications of these methods for the resulting uncertainty?" The choice of these methods does not influence the result in the fuzzy EP-based methodology (Subsection 4.1.4) so long as enough values of the coefficients are taken as are required for the determination of the maximum and the minimum. This is because, in the EP-based method, it is not the probability that counts. Since we are concerned only to find the maximum and minimum of the output, only one set of coefficients that results in maximum and minimum is sufficient. On the other hand, in the MC based method (Subsection 4.1.5), it is the number of occurrences of such sets that counts. If such sets are very few, they will not have significance for the output uncertainty. Therefore, the choice of the methods for generating the coefficients has a direct influence in the MC based methodology. The normalisation method is perhaps the most rational one for use with the MC-based methodology, as it gives a high probability of generating mean values.

4.2 Improved first–order second moment method

The FOSM method (see Subsection 3.1.2) is one of the most widely used methods for uncertainty estimation in model results due to uncertainty of the parameters. Some successful applications of FOSM in water related problems are presented by Lee and Mays (1986), Melching (1992 & 1995) and Tung and Mays (1981). Several advantages of the FOSM method are that (i) it does not require the knowledge of input parameter distributions, (ii) it is computationally efficient and (iii) it provides a measure of the sensitivity of the model outputs to the input random variables (Tyagi and Haan, 2001). Like any other method, this method suffers from a number of limitations that arise from the first-order approximation. To improve the accuracy of the uncertainty estimation by this method, several improvements or modifications have been proposed in various contexts (Kunstmann et al., 2002; Melching and Yoon, 1996; Tyagi and Haan, 2001).

The research result presented here consists in improving the uncertainty estimates of the FOSM method. This is an original contribution of the present research and is called the Improved FOSM method (Maskey and Guinot, 2002 & 2003). The method is applied to an operational flood forecasting model of the Loire River (France). The results of the improved method are compared with that of FOSM and MC methods and are presented in Chapter 6.

4.2.1 Practical implementation of FOSM method

Before introducing the improved method, the practical implementation of the conventional FOSM method is discussed. In Chapter 3, a function y is introduced that relates several random input variables $X_1, \ldots, X_n$ to an output random variable Y, i.e.

$$Y = y(X_1, \ldots, X_n) \tag{4.15}$$

In most practical applications, there exists no analytical expression for y. This is because y is normally a numerical result given by a simulation model. Consequently, the derivatives $\partial y / \partial X_i$ (Equation (3.8), Chapter 3) cannot be determined analytically. This is the case, for instance, when y is the output variable of a distributed, physically-based model. The classical solution to this problem consists of approximating y with a linear function f:

$$f(X_1, \ldots, X_n) = y(\bar{x}_1, \ldots, \bar{x}_n) + \sum_{i=1}^{n} (X_i - \bar{x}_i) \frac{\partial f}{\partial X_i} \tag{4.16}$$

The derivatives $\partial f / \partial X_i$ are estimated using finite differences around the mean values:

$$\frac{\partial f}{\partial X_i} = \frac{y(\bar{x}_1, \ldots, \bar{x}_{i-1}, x_{i,2}, \bar{x}_{i+1}, \ldots, \bar{x}_n) - y(\bar{x}_1, \ldots, \bar{x}_{i-1}, x_{i,1}, \bar{x}_{i+1}, \ldots, \bar{x}_n)}{x_{i,2} - x_{i,1}} \tag{4.17}$$

where $x_{i,1}$ and $x_{i,2}$ are two different values of X_i taken around the mean value. The most commonly used options are the following (see Fig. 4.6):

- The forward method uses $x_{i,1} = \bar{x}_i$ and $x_{i,2} = \bar{x}_i + \varepsilon$,

- The backward method uses $x_{i,1} = \bar{x}_i - \varepsilon$ and $x_{i,2} = \bar{x}_i$,

- The centred method uses $x_{i,1} = \bar{x}_i - \varepsilon$ and $x_{i,2} = \bar{x}_i + \varepsilon$

where ε is a perturbation, usually taken as a fraction of the standard deviation of X_i. The mean and the variance of Y can be obtained by replacing y with f in Equations (3.5) and (3.8) presented in Chapter 3.

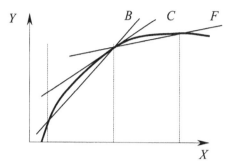

Figure 4.6. Three options for derivative assessment. The bold line represent the original function y and the straight lines represent the approximate functions f obtained using the backward (B), centred (C) and forward (F) difference methods.

4.2.2 Problems attached to FOSM method

Several theoretical and/or conceptual problems in the application of the FOSM method are:

- The function f aims to approximate a function y that is nonlinear in most cases. Therefore, the computed value $f(X_1,...,X_n)$ may depart substantially from the actual value Y when the standard deviation of the input variable is large compared to the interval over which the function f is assumed to be linear.

- When the function y is highly nonlinear, the method poses problems in that (i) the backward, centred or forward estimates often result in very different computed values, and (ii) the result is very sensitive to the size of the perturbation ε.

- Eventually, a problem that has not been discussed widely in literature occurs when the mean value of the input variable is very close to a local or global extremum of the function. In this case, the computed standard deviation (and consequently the uncertainty in the variable) given by the FOSM method may be very different from the real value. Figure 4.7 illustrates such a situation. In this example, the function y has an extremum close to the average $\bar{x}$. Assume now that the range of the uncertainty in X is given by the interval $[x_1, x_2]$. This results in the range of uncertainty $[y_1, y_2]$ for Y. However, the FOSM method, which consists of using the tangent of the linearised function f around $\bar{x}$, would give an interval $[f_1, f_2]$ different from the real one $[y_1, y_2]$. In particular, f_2 is larger than y_2 because the real function $Y(X)$ takes its maximum within the interval $[x_1, x_2]$. Moreover, f_1 is larger than y_1 because the function $Y(X)$ is convex.

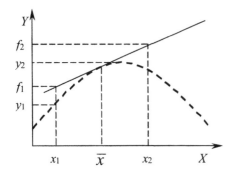

Figure 4.7. An example where the classical FOSM fails to identify the correct range of variation of the output variable. The upper bound f_2 of the linearised function f is larger than the upper bound y_2 of the real function y because of the extremum in the function y. The lower bound f_1 of f is lower than the lower bound y_1 of y because of the convex nature of the real function.

4.2.3 Principle of the improved method

One possible way to improve the accuracy of the approximation is to use higher-order terms from the Taylor series expansion. Unfortunately, the algebraic complexity increases rapidly with the inclusion of higher-order terms. Moreover, the inclusion of higher order terms also requires information on higher-order central moments, such as skewness and kurtosis (Haldar and Mahadevan, 2000a). In practice, these moments are difficult to assess due to the lack of data.

The Improved First-Order Second Moment (IFOSM) method presented here aims at correcting the undesirable behaviour of the FOSM method near extrema (which is the result of the linearisation) without making the computation too complicated. The principle of the IFOSM method is based on the following premises:

1. The input-output function y is approximated using a parabolic function, that is, a second-degree approximation.

2. The unknown PDFs of the input variables are replaced by equivalent uniform or triangular density functions, which are derived from the given means and standard deviations of the input variables.

3. The uncertainty is evaluated for the whole range of the variable defined by its equivalent PDF, instead of using only the point estimate about the mean value as in the FOSM method.

4.2.4 Mathematical derivation

For simplicity, the mathematical derivation of the IFOSM method is carried out for the function of a single input variable X. The total variance of the output variable Y due to the variances of all input variables can be computed for multiple input variables using Equation (3.8), assuming that the inputs are independent. Let the function of y in $Y = y(X)$ be approximated by a parabolic function (Fig. 4.8), that is:

$$Y = f(X) = aX^2 + bX + c \qquad (4.18)$$

If there are multiple input variables, say n input variables, there will be n number of equations like Equation (4.18). In which case, the coefficients a, b, and c of each of them are the functions of the mean values of the remaining variables.

It is assumed here that the problem is solved numerically by applying a perturbation on the input variable. Three points x_C, x_B and x_F are defined as follows

$$x_B = x_C - \varepsilon \qquad (4.19)$$

$$x_F = x_C + \varepsilon \qquad (4.20)$$

$$x_C = \overline{x} \qquad (4.21)$$

where $\overline{x}$ is equal to the mean value of the variable.

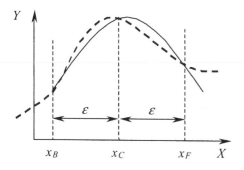

Figure 4.8. Approximation of the real function y (bold, dashed line) with a parabolic function f (thin line) using the three points x_B, x_C and x_F.

The first step in the method consists of identifying the coefficients a, b and c. This is achieved by requiring the function f to take the same values as y at the points x_B, x_C and x_F. This yields the following equalities:

$$a = \frac{y_B - 2y_C + y_F}{2\varepsilon^2} \tag{4.22}$$

$$b = -\frac{(2x_C + \varepsilon)y_B - 4x_C y_C + (2x_C - \varepsilon)y_F}{2\varepsilon^2} \tag{4.23}$$

$$c = \frac{(x_C + \varepsilon)x_C y_B - 2(x_C^2 - \varepsilon^2)y_C + (x_C - \varepsilon)x_C y_F}{2\varepsilon^2} \tag{4.24}$$

where $y_B = y(x_B)$, $y_C = y(x_C)$ and $y_F = y(x_F)$. The function f is then determined uniquely. If there are multiple input variables, y_B, y_C and y_F are determined as follows:

$$\left. \begin{array}{l} y_B = y(\overline{x}_1,...,x_{i,B},...,\overline{x}_n) \\ y_C = y(\overline{x}_1,...,x_{i,C},...,\overline{x}_n) \\ y_F = y(\overline{x}_1,...,x_{i,F},...,\overline{x}_n) \end{array} \right\} \tag{4.25}$$

The second step of the method consists of assuming the probability density function p_X for X and using f as a substitute for y in Equations (3.2) and (3.3) to compute the mean and variance of Y. Two examples are provided for uniform and triangular forms of p_X.

Case of a uniform probability density function

Assume that $p_X(x)$ is uniform. Denoting by L the half length of the support of the PDF and its variance by σ_X, it is easy to check that:

$$p_X(x) = \frac{1}{2L} \tag{4.26}$$

$$L = \sqrt{3}\sigma_X \tag{4.27}$$

Substituting Equations (4.26) and (4.27) into Equations (3.2) and (3.3) yields:

$$\begin{aligned} E(Y) &= \int_{X_C-L}^{X_C+L} f(X)p_X(x)dx \\ &= \frac{1}{2L}\int_{X_C-L}^{X_C+L}(aX^2 + bX + c)dx \\ &= a\left(X_C^2 + \frac{L^2}{3}\right) + bX_C + c \end{aligned} \tag{4.28}$$

$$Var(Y) = \sigma_Y^2 = \int_{X_C-L}^{X_C+L} [Y - E(Y)]^2 \, p_X(x) dx$$

$$= \frac{1}{2L} \int_{X_C-L}^{X_C+L} [Y - E(Y)]^2 \, dx \tag{4.29}$$

$$= \frac{L^2}{45} \left(60a^2 X_C^2 + 4a^2 L^2 + 60abX_C + 15b^2 \right)$$

Case of a triangular probability density function

The analytical expression for a symmetrical PDF can be written as:

$$p_X(x) = \begin{cases} \dfrac{L - |x - X_C|}{L^2} & \text{for } X_C - L \leq x \leq X_C + L \\ 0 & \text{otherwise} \end{cases} \tag{4.30}$$

where the half length of the triangular PDF, L, is given by

$$L = \sqrt{6} \sigma_X \tag{4.31}$$

Again, substituting Equations (4.30) and (4.31) into Equations (3.2) and (3.3) the mean and variance of the output Y can be computed as

$$E(Y) = \int_{X_C-L}^{X_C+L} f(X) p_X(x) dx = a \left(X_C^2 + \frac{L^2}{6} \right) + bX_C + c \tag{4.32}$$

$$Var(Y) = \sigma_Y^2 = \int_{X_C-L}^{X_C+L} [Y - E(Y)]^2 \, p_X(x) dx$$

$$= \frac{L^2}{180} \left(120a^2 X_C^2 + 7a^2 L^2 + 120abX_C + 30b^2 \right) \tag{4.33}$$

4.3 Qualitative scales for uncertainty interpretation

A qualitative method for uncertainty assessment based on expert judgement and fuzzy set theory has been presented in Chapter 3 (Subsection 3.2.2). In this method the uncertainty in each input for quality and importance is represented by linguistic variables and is defined by a membership function. In contrast with the MF used in the fuzzy Extension Principle-based method (Subsection 3.2.1), the base value of the MF in the expert judgement-based qualitative method is defined in terms of an arbitrary unit. Consequently, the estimated MF of the output also has its base value in an arbitrary unit. It, therefore, does not directly show the uncertainty bounds (upper and lower) for the output in the unit for the quantitative interpretation of the output variable. The result of this method is therefore qualitative and hence suited for qualitative interpretation.

The present research proposed a Qualitative Uncertainty Scale (QUS) in which the estimated qualitative result can be measured (Maskey et al., 2002a). The procedures of deriving the QUS are explained in Subsection 4.3.1. An example is presented in Subsection 4.3.2 to show how the estimated uncertainty is represented on the QUS.

4.3.1 Derivation of Qualitative Uncertainty Scales

The QUS scale is derived using the concept of the best-case and the worst-case scenarios. The best-case scenario corresponds to the case where the Qualities of all the sub-parameters are assigned to the highest level and the worst-case scenario corresponds to the case where all the Qualities are assigned to the lowest level. For example, if five variables *very bad, bad, acceptable, good* and *very good* are used, all qualities will be *very good* for the best-case and *very bad* for the worst-case. In both cases the Importance of the parameters are taken from the experts' evaluation. That is, from Equation (3.25), we obtain

$$U_O(\text{best}) = \overset{\text{fuzzy sum}}{\underset{\text{all } j}{\sum}} \Phi'_j(\text{best})(\times)\Psi_j \tag{4.34}$$

$$U_O(\text{worst}) = \overset{\text{fuzzy sum}}{\underset{\text{all } j}{\sum}} \Phi'_j(\text{worst})(\times)\Psi_j \tag{4.35}$$

The estimated uncertainty values corresponding to the best- and the worst-cases are then used to represent the two extremes (the highest and the lowest levels, respectively) on the qualitative scales. The scales between the two extremes are derived by linear interpolations at different membership levels. The interpolations are performed separately for the lower and upper bounds. Let each division of the scale be represented by a fuzzy number U_i ($i = 1,\ldots, n$; n is the number of divisions (levels) on the scales). Then the lower bound and upper bound of U_i at each belief level α is given by

$$U_{i,LB}^{(\alpha)} = U_{1,LB}^{(\alpha)} + \frac{U_{n,LB}^{(\alpha)} - U_{1,LB}^{(\alpha)}}{(n-1)}(i-1) \tag{4.36}$$

$$U_{i,UB}^{(\alpha)} = U_{1,UB}^{(\alpha)} + \frac{U_{n,UB}^{(\alpha)} - U_{1,UB}^{(\alpha)}}{(n-1)}(i-1) \tag{4.37}$$

where

$$\left.\begin{aligned} U_{1,LB}^{(\alpha)} &= U_{LB}^{(\alpha)}(\text{best}) \\ U_{1,UB}^{(\alpha)} &= U_{UB}^{(\alpha)}(\text{best}) \end{aligned}\right\} \tag{4.38}$$

$$U_{n,LB}^{(\alpha)} = U_{LB}^{(\alpha)}(\text{worst}) \Big\}$$
$$U_{n,UB}^{(\alpha)} = U_{UB}^{(\alpha)}(\text{worst}) \Big\}$$
$$(4.39)$$

The scales derived using Equations (4.34) through (4.39) are in fuzzy scales because the regions of different uncertainty levels are represented by fuzzy variables. The fuzzy scales are transformed to crisp scales such that the boundaries between the two scales are non-fuzzy. In this transformation from fuzzy to crisp scales, the abscissa passing through the point of intersection of two consecutive levels on fuzzy scales separates the two consecutive levels on the crisp scales. An example of fuzzy and crisp scales is shown in Figure 4.9. The QUS in this example consist of five divisions: Very Small Uncertainty (VSU), Small Uncertainty (SU), Moderate Uncertainty (MU), Large Uncertainty (LU) and Very Large Uncertainty (VLU). The VSU and VLU correspond to the best- and worst-case scenarios, respectively. In the figure, the boundaries of fuzzy scales are shown in thin-dark and thick-grey lines and the boundaries of crisp scales in thick-dark vertical lines. Note that the QUS varies case to case. The scales shown here is for the example presented in Subsection 4.3.2.

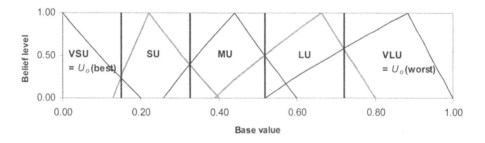

Figure 4.9. An example of fuzzy and crisp qualitative scales. The fuzzy scales are represented in thin-dark and thick-grey lines and the crisp scales are represented in thick-dark likes.

4.3.2 Presentation of uncertainty in Qualitative Uncertainty Scale

In order to illustrate how the computed uncertainty is measured on the QUS, a model of two input uncertain (fuzzy) variables A and B are considered. Let the output of the model is C (also fuzzy), such that $C = f(A, B)$. Qualitative assessments about the quality and importance of the inputs A and B are arbitrarily represented by membership functions as shown in Figure 4.10(a, b). The estimated uncertainty in C (the method is explained in Subsection 3.2.2) is shown in Figure 4.11.

The estimated uncertainty is then presented in the QUS, which is derived according to Subsection 4.3.1 (Fig. 4.12 (a, b)). Both fuzzy and crisp scales are shown. To present the uncertainty on the crisp scales, the fuzzy representation of the estimated uncertainty in C is transformed to a crisp representation by the so-called defuzzification. Various methods of defuzzification are available. Some commonly used methods are presented in Appendix I. For the application in this example the *centre-of-area* method is used.

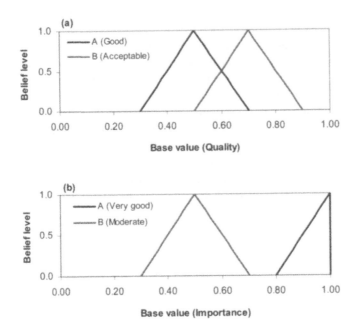

Figure 4.10. Qualitative representation of uncertainty in two inputs A and B: (a) Quality, (b) Importance.

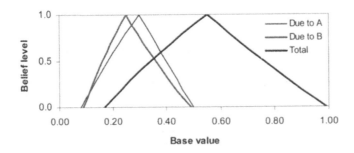

Figure 4.11. Estimated uncertainty in the output $C = f(A, B)$.

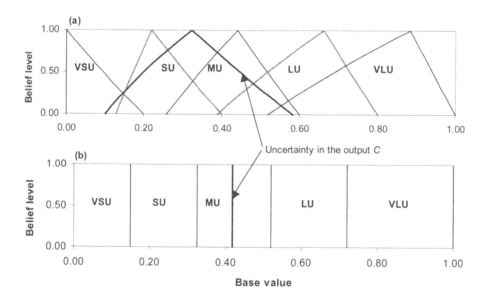

Figure 4.12. Example of presenting uncertainty in (a) fuzzy and (b) crisp qualitative uncertainty scales. The estimated uncertainty in the output C is shown in a thick-dark line.

4.4 Towards hybrid techniques of modelling uncertainty

A hybrid technique of uncertainty modelling is defined here as a method that makes use of both probabilistic and possibilistic approaches together. Although there are some analogies, the concepts that lead to the formulation of probability theory and possibility theory are very different. The former considers uncertainty as random, whereas the latter assumes uncertainty due to imprecision and vagueness (Ross, 1995). In order to analyse the situations where both types of uncertainty are present two types of problem are distinguished: these are referred to as Type I Problem and Type II Problem.

4.4.1 Type I and Type II Problems

In many situations, an uncertain parameter may contain components of two major types of uncertainty: one due to randomness and the other due to imprecision and vagueness (or fuzziness). Probability theory and fuzzy set theory (or possibility theory) are used to represent these types of uncertainty, respectively. In the presence of both types of uncertainty, representing uncertainty by either one of the two approaches alone is considered incomplete. This type of problem is referred to here as a "Type I Problem". Various concepts have been emerged to deal with this type of problem. In-depth coverage of this topic is beyond the scope of this thesis. Kaufmann and Gupta (1991) presented some discussions on the issue. Two particular concepts have been presented in Chapter 3. They are the concept of *fuzzy probability* by Zadeh (1968 and 1984) and the concept of *fuzzy-random variable* by Ayyub and Chao (1998).

For a given system, there may exist some parameters for which enough data are available to characterise their uncertainty probabilistically (using PDFs) by statistical methods. In contrast there may remain some uncertain parameters with very little or no data to characterise their uncertainty probabilistically. This is a very common problem in uncertainty analysis in practice. What is usually done is to exclude such parameters (with no data for uncertainty characterisation) from the analysis, or assume their PDFs based on judgements. This latter set of parameters may be more suitably characterised for uncertainty by means of the possibilistic approach using the possibility distributions or fuzzy membership functions. This then gives rise to a system with two sets of uncertain parameters, one represented by PDFs and the other represented by MFs. This type of situation is referred to here as "Type II Problem". The Type II Problem is more practical but, ironically, more difficult to deal with.

An ideal solution to the Type II Problem could be to separate the uncertain parameters into probabilistic (defined by PDFs) and possibilistic (defined by MFs), and transform from one type to another as appropriate. Some of the methods for such transformations are discussed in Chapter 3. These and other transformation methods reported in the literature are neither fully rational nor are free from criticism.

4.4.2　Operations on random and fuzzy variables

Uncertain variables represented by PDFs and MFs are commonly referred to as random and fuzzy variables, respectively. Note that random and fuzzy variables are sometimes also referred to as random and fuzzy numbers (Kaufmann and Gupta, 1991). The methods of operations between random-random (R-R) and fuzzy-fuzzy (F-F) variables are very different. The major obstacle in modelling the uncertainty of a system consisting of random and fuzzy variables comes from these differences.

The Monte Carlo (MC) method and the fuzzy Extension Principle (EP) are the two standard methods for the operations on random variables and fuzzy variables, respectively. Some publications that address the issue of differences in MC and EP are Chen et al. (1998), Dubois and Prade (1991), Ferson and Kuhn (1994), Ferson and Ginzburg (1995), Fishwick, (1991), Guyonnet et al. (1999), Kaufmann and Gupta (1991) and Schulz and Huwe (1997). Some discussion on the issue is also presented in Subsection 3.6.1 of this thesis. Two notable differences between the MC and the EP arise from (i) the correlation between the variables, and (ii) the nature of the function (monotonic or non-monotonic). In the MC method, the correlations can be incorporated. In the EP, on the other hand, there is no provision (with the present state-of-the-art) for the treatment of correlations between variables. The EP, however, can be greatly simplified if the function is monotonic.

These differences can be illustrated by taking a simple example of the addition of two independent random variables defined by uniform PDFs and two fuzzy variables defined by constant MFs (Schulz and Huwe, 1997). As shown in Figure 4.13, the summation of two uniformly distributed independent random variables results in a triangular distribution, whereas the sum of two fuzzy variables with constant MFs results in a constant MF (Fig. 4.14).

In essence, the operation on fuzzy variables by the EP is a deterministic method whereas the operation on random variables by the MC method is non-deterministic. As illustrated later in this section, if two variables are functionally dependent on each other, in certain cases the operation on the random variables also happens to be deterministic. In which case operations on random variables and on fuzzy variables with monotonic functions are similar.

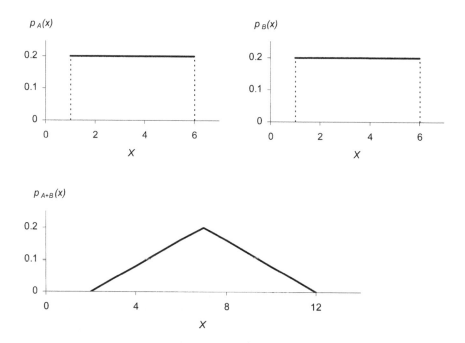

Figure 4.13. Summation of two independent random variables with uniform PDFs (derived by the Monte Carlo method).

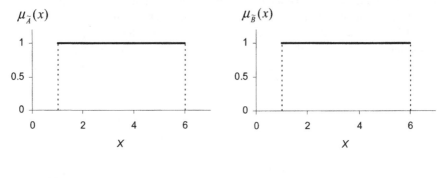

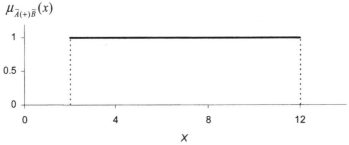

Figure 4.14. Summation of two fuzzy variables with constant MFs (derived by the Extension Principle).

In order to illustrate this, consider two random variables X and Y and two fuzzy variables $\widetilde{X}$ and $\widetilde{Y}$. Let Z and $\widetilde{Z}$ be random and fuzzy variables, respectively, such that

$$Z = X + Y \tag{4.40}$$

$$\widetilde{Z} = \widetilde{X}(+)\widetilde{Y} \tag{4.41}$$

Note that in Equation (4.41) for the addition between two fuzzy variables, the addition symbol is enclosed by parentheses to distinguish it from the addition between random variables. Four basic operations on fuzzy variables are presented in Appendix I. For any value of α-cuts ($\alpha = [0, 1]$), the addition of two fuzzy variables (Fig. 4.15), that is $\widetilde{Z}$ in Equation (4.41) is given by

$$\left.\begin{aligned} z_{LB}^{(\alpha)} &= x_{LB}^{(\alpha)} + y_{LB}^{(\alpha)} \\ z_{UB}^{(\alpha)} &= x_{UB}^{(\alpha)} + y_{UB}^{(\alpha)} \end{aligned}\right\} \tag{4.42}$$

Equation (4.42) can also be written as following

$$\left.\begin{array}{l} \mu^{-1}_{\tilde{Z}=\tilde{X}(+)\tilde{Y},L}(\alpha)=\mu^{-1}_{\tilde{X},L}(\alpha)+\mu^{-1}_{\tilde{Y},L}(\alpha) \\ \mu^{-1}_{\tilde{Z}=\tilde{X}(+)\tilde{Y},R}(\alpha)=\mu^{-1}_{\tilde{X},R}(\alpha)+\mu^{-1}_{\tilde{Y},R}(\alpha) \end{array}\right\}$$

(4.43)

where the subscripts L and R represent LB (lower bound) and UB (upper bound), respectively.

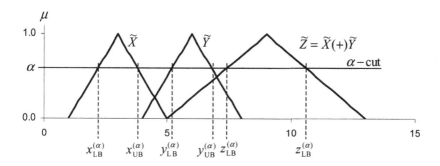

Figure 4.15. Addition of two fuzzy variables.

Similarly, the three other operations (subtraction, multiply and division) on fuzzy variables are given by

$$\left.\begin{array}{l} \mu^{-1}_{\tilde{Z}=\tilde{X}(-)\tilde{Y},L}(\alpha)=\mu^{-1}_{\tilde{X},L}(\alpha)-\mu^{-1}_{\tilde{Y},R}(\alpha) \\ \mu^{-1}_{\tilde{Z}=\tilde{X}(-)\tilde{Y},R}(\alpha)=\mu^{-1}_{\tilde{X},R}(\alpha)-\mu^{-1}_{\tilde{Y},L}(\alpha) \end{array}\right\}$$

(4.44)

$$\left.\begin{array}{l} \mu^{-1}_{\tilde{Z}=\tilde{X}(\times)\tilde{Y},L}(\alpha)=\mu^{-1}_{\tilde{X},L}(\alpha)\mu^{-1}_{\tilde{Y},L}(\alpha) \\ \mu^{-1}_{\tilde{Z}=\tilde{X}(\times)\tilde{Y},R}(\alpha)=\mu^{-1}_{\tilde{X},R}(\alpha)\mu^{-1}_{\tilde{Y},R}(\alpha) \end{array}\right\}$$

(4.45)

$$\left.\begin{array}{l} \mu^{-1}_{\tilde{Z}=\tilde{X}(:)\tilde{Y},L}(\alpha)=\mu^{-1}_{\tilde{X},L}(\alpha)/\mu^{-1}_{\tilde{Y},R}(\alpha) \\ \mu^{-1}_{\tilde{Z}=\tilde{X}(:)\tilde{Y},R}(\alpha)=\mu^{-1}_{\tilde{X},R}(\alpha)/\mu^{-1}_{\tilde{Y},L}(\alpha) \end{array}\right\}$$

(4.46)

In the case of random variables, suppose two random variables are related by a function g such that

$$Y=g(X)$$

(4.47)

If g is a monotonically increasing function of x with a unique inverse g^{-1}, then the CDF of Y can be related to the CDF of X by the relationship (Ang and Tang, 1975, p. 170):

$$P_Y(y)=P_X[g^{-1}(y)]$$

(4.48)

where g^{-1} is the inverse function of g.

The summation Z of X and Y, which are related by the function g, is given by

$$\begin{aligned} Z &= X + Y \\ &= X + g(X) \\ &= f(X) \end{aligned} \qquad (4.49)$$

Now, if the function f is also monotonically increasing function of x, then using Equation (4.48), the CDF of Z can be expressed in terms of the CDF of X as follows:

$$P_Z(z) = P_X[f^{-1}(z)] \qquad (4.50)$$

The simplest relationship between X and Y that keeps the function f monotonically increasing with x is a linear relationship, i.e. $Y = a + bX$, with $b \geq 0$.

From Equation (4.50), for any value $\beta = [0, 1]$, it follows

$$P_Z^{-1}(\beta) = z \qquad (4.51)$$

and

$$P_X^{-1}(\beta) = f^{-1}(z) \qquad (4.52)$$

Since z is related to x as given by Equation (4.49), we obtain

$$\begin{aligned} z &= P_X^{-1}(\beta) + g(P_X^{-1}(\beta)) \\ &= P_X^{-1}(\beta) + P_Y^{-1}(\beta) \end{aligned} \qquad (4.53)$$

Now, from Equations (4.51) and (4.53), we obtain

$$P_{Z=X+Y}^{-1}(\beta) = P_X^{-1}(\beta) + P_Y^{-1}(\beta) \qquad (4.54)$$

Similarly, the following can be derived for multiplication between X and Y ($Z = XY$, $X = g(Y)$), if Z is monotonically increasing as follows:

$$P_{Z=X\times Y}^{-1}(\beta) = P_X^{-1}(\beta) P_Y^{-1}(\beta) \qquad (4.55)$$

Now, referring to Equation (4.49), if the function f is monotonically decreasing with x, then the CFD of Z is given by (Ang and Tang, 1975, p. 171):

$$P_Z(z) = 1 - P_X[f^{-1}(z)] \qquad (4.56)$$

For a linear relationship between X and Y (i.e. $Y = a + bX$), with $b \geq 0$, the function f becomes monotonically decreasing with x if $Z = X - Y$. Then from Equation (4.56), it can be shown (for subtraction) that

$$P_{Z=X-Y}^{-1}(\beta) = P_X^{-1}(1-\beta) - P_Y^{-1}(1-\beta) \tag{4.57}$$

Similarly, for division, i.e. $Z = X / Y$, with z monotonically decreasing with x and $y \neq 0$, we obtain:

$$P_{Z=X:Y}^{-1}(\beta) = P_X^{-1}(1-\beta) / P_Y^{-1}(1-\beta) \tag{4.58}$$

It can be observed that Equations (4.43) and (4.45) are similar to Equations (4.54) and (4.55), respectively. These two sets of equations are for addition and multiplication. This suggests that the addition and multiplication of two functionally dependent random variables (if their sum and product are monotonically increasing) can be made equivalent to the corresponding operations on fuzzy variables, by applying an appropriate transformation between their CDF and MF.

The operations are not similar in the case of subtraction and division. The difference is due to the reversed order of the appearance of the lower and upper bound values of the operating variables in fuzzy subtraction and division (Equations (4.44) and (4.46)).

4.4.3 Probability–fuzzy transformations

Let us define two transformations: T_1 from probability (in the form of a CDF) to fuzzy (in the form of a MF) and T_2 from MF to CDF for two fuzzy operations $(\circ) \in [(+),(\times)]$ and two random operations $\circ \in [+,\times]$. The transformation T_1 is such that

$$\widetilde{X}(\circ)\widetilde{Y} = T_1(X \circ Y) \tag{4.59}$$

where

$$\left. \begin{aligned} \widetilde{X} &= T_1(X) \\ \widetilde{Y} &= T_1(Y) \end{aligned} \right\} \tag{4.60}$$

Similarly the transformation T_2 is such that

$$X \circ Y = T_2(\widetilde{X}(\circ)\widetilde{Y}) \tag{4.61}$$

where

$$\left.\begin{array}{l} X = T_2(\tilde{X}) \\ Y = T_2(\tilde{Y}) \end{array}\right\} \tag{4.62}$$

A linear transformation is proposed. The transformation from $P(x)$ to $\mu(x)$, that is T_1, is the following:

$$\mu(x) = \begin{cases} \dfrac{1}{P_0} P(x) & \text{for} & x \le x_0 \\[2ex] \dfrac{1 - P(x)}{1 - P_0} & \text{for} & x \ge x_0 \end{cases} \tag{4.63}$$

The transformation T_2, i.e. from $\mu(x)$ to $P(x)$, is then given by:

$$P(x) = \begin{cases} P_0 \mu(x) & \text{for} & x \le x_0 \\ 1 - (1 - P_0)\mu(x) & \text{for} & x \ge x_0 \end{cases} \tag{4.64}$$

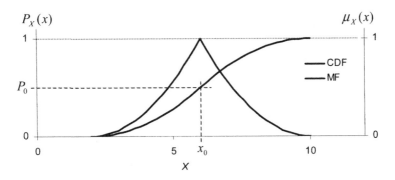

Figure 4.16. Transformation between CDF and MF.

As shown in Figure 4.16, P_0 is the value of P_X at $x = x_0$. This results in $\mu(x = x_0) = 1$, which is normally interpreted as the most credible value for the MF. Therefore, x_0 should preferably be one of the central values of the PDF, i.e. mean, median or mode. If the x_0 is taken as the median of the distribution, P_0 becomes 0.5 and the transformations will take the form

$$\mu(x) = \begin{cases} 2P(x) & \text{for} & x \le x_0 \\ 2(1 - P(x)) & \text{for} & x \ge x_0 \end{cases} \tag{4.65}$$

$$P(x) = \begin{cases} \dfrac{\mu(x)}{2} & \text{for} & x \le x_0 \\[2ex] 1 - \dfrac{\mu(x)}{2} & \text{for} & x \ge x_0 \end{cases} \tag{4.66}$$

Therefore, with these transformations Equations (4.43) and (4.45) are equivalent to Equations (4.54) and (4.55) respectively, when

$$\alpha = \begin{cases} 2\beta & \text{for} \quad 0 \le \beta \le 0.5 \\ 2(1-\beta) & \text{for} \quad 0.5 \le \beta \le 1 \end{cases} \tag{4.67}$$

The equivalency between Equations (4.44) and (4.57) (subtraction operation) and between Equations (4.46) and (4.58) (division operation) can be established, if Equations (4.57) and (4.58) are modified as follows:

$$P^{-1}_{(Z=X-Y)^*}(\beta) = P_X^{-1}(1-\beta') - P_Y^{-1}(1-\beta) \tag{4.68}$$

$$P^{-1}_{(Z=X:Y)^*}(\beta) = P_X^{-1}(1-\beta') / P_Y^{-1}(1-\beta) \tag{4.69}$$

where $\beta' = [0,1]$. The superscript (*) is used in Equations (4.68) and (4.69) to indicate that they are modified operations. For the proposed transformation (Equation (4.65)), it is easy to check that $\beta' = 1 - \beta$. This makes Equations (4.44) and (4.46) equivalent to Equations (4.68) and (4.69), respectively for the α and β related by Equation (4.67).

4.4.4 Application example

In order to illustrate the application of the CDF-MF transformations proposed above, the uniform flow formula of the form given by Equation (4.70) is used, i.e.

$$Q = \frac{B H^{\frac{5}{3}} S^{\frac{1}{2}}}{N} \tag{4.70}$$

where Q is the discharge (m³/s), B and H are the channel width (m) and water depth (m), respectively, S is the water surface slope (m/m) and N is the Manning's roughness coefficient (m$^{-1/3}$s). This form of the uniform flow formula assumes a wide rectangular channel with $P = B + 2H \approx B$.

This illustration consists of the following steps:

1. The B, H, S and N are assumed as uncertain input variables, each defined by a bounded normal PDF. The CDFs of these inputs derived from their PDFs are given in Figure 4.17.

2. The uncertainty in the output Q due to the uncertainty in the inputs is estimated by applying a modified MC simulation (also referred here as a random simulation). As required by Equation (4.55) (for multiplication) and Equation (4.69) (for division), the modified MC simulations are performed in such a way that in each simulation the same value of a random number (β) is

used to sample values of B, H and S and a random number $1 - \beta$ is used to sample a value of N. The output uncertainty in the form of the CDF is shown in Figure 4.18.

3. The CDFs of the inputs are transformed to MFs using the transformation given by Equation (4.65). The transformed MFs are shown in Figure 4.17.

4. The uncertainty in the output Q is estimated due to the input uncertainty in the form of MF (transformed from CDF) by applying the fuzzy EP using the α-cut method (also referred here as a fuzzy simulation). The output uncertainty in the form of the MF is shown in Figure 4.19.

5. The output CDF obtained in Step 2 (from random simulation) is transformed to a MF by applying the transformation defined by Equation (4.65) and compared with the output MF obtained in Step 4 (from fuzzy simulation) as shown in Figure 4.18.

6. Similarly, the output MF obtained in Step 4 is transformed to a CDF by applying the transformation defined by Equation (4.66) and compared with the output CDF obtained in Step 2 as shown in Figure 4.19.

The comparisons show that the results are in perfect agreement. An implication of this result is that it provides an alternative method for the evaluation of the Extension Principle for a monotonic function without using the α-cut method.

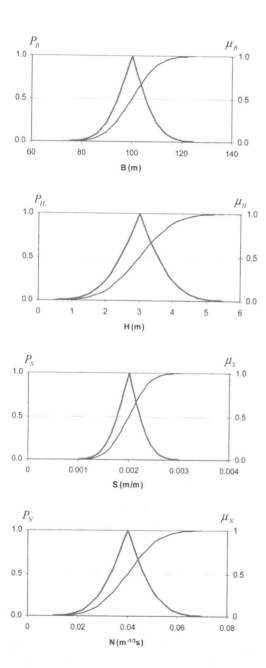

Figure 4.17. Uncertainty in inputs *B*, *H*, *S* and *N*: as CDFs (dark line) derived from bounded normal PDFs, and as MF (grey line) transformed from the CDFs.

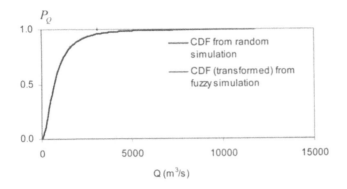

Figure 4.18. Uncertainty in the output Q in the form of a CDF: one obtained directly from random simulation (by modified MC simulation) and the other transformed from a MF, which is obtained from fuzzy simulation (by EP).

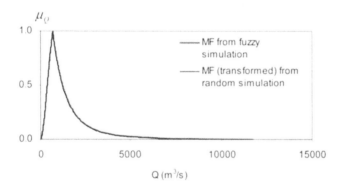

Figure 4.19. Uncertainty in the output Q in the form of a MF: one obtained directly from fuzzy simulation (by EP) and the other transformed from a CDF, which is obtained from random simulation (by modified MC simulation).

4.4.5 Concluding remarks

It is proved here that the addition and multiplication of two fuzzy variables by the Extension Principle using α-cut is similar to corresponding operations between two functionally dependent random variables for some specific conditions. The dependency should be such that the sum and the product are monotonically increasing. This means that the addition and multiplication operations on such random variables can be performed by the EP using a transformation between random and fuzzy variables. These transformations are applicable for subtraction and division, only if the operations on random variables are modified (as given by Equations (4.68) and (4.69)). An implication of the results of the transformations applied to the uniform flow formula is that it provides an alternative method for the evaluation of the Extension Principle for a monotonic function without using the α-cut method. It appears that the direct implication of this finding is very limited. However, it is believed that these findings are important for further research in probability–possibility transformations and hybrid techniques of uncertainty modelling.

It should be noticed that the transformation proposed here is solely based on the characteristics of the operations on random and fuzzy variables, and it guarantees neither the probability/possibility consistency (Zadah, 1978) nor the uncertainty measures (Klir, 1992).

Chapter 5

APPLICATION: FLOOD FORECASTING MODEL FOR KLODZKO CATCHMENT (POLAND)

Summary of Chapter 5

This chapter presents the application of the uncertainty assessment methodology using temporal disaggregation (developed in Section 4.1) to rainfall time series. The methodology uses temporal disaggregation of the time series into subperiods and uses the fuzzy Extension Principle for the propagation of the uncertainty through the forecasting model. Genetic algorithms are used to determine the maxima and minima, which is an essential part of the Extension Principle. A rainfall-runoff model of the Klodzko catchment (Poland) built with HEC-1 and HEC-HMS is used for the application. The results show that the output uncertainty arising from the uncertain temporal distribution of the precipitation can be significantly dominant over the uncertainty arising from the uncertainty in the magnitude of the precipitation. The results also show the potential applicability of genetic algorithms in combination with the Extension Principle for uncertainty assessment. The description of the Klodzko catchment and the flood forecasting model are presented in Section 5.1. The implementation of the methodology is described in Section 5.2. The principle and the versions of genetic algorithms used in the present application are presented in Section 5.3. The results of the application presented in Section 5.4 are analysed as follows: (i) uncertainty in the forecast discharges with reconstructed precipitation (with disaggregation) and with uniform precipitation (without disaggregation), (ii) comparison of results with 3 and 6 subperiods, (iii) influence of time series correlation between subbasins, and (iv) influence of the type of genetic algorithms used for the determination of the minimum and maximum. The results are presented in the form of upper and lower bounds on the forecasts and in the form of a complete membership function for some selected results. The conclusions and discussion are presented in Section 5.5.

5.1 Klodzko catchment flood forecasitng model

A methodology for the treatment of uncertainty using temporal disaggregation is presented in Chapter 4 (Section 4.1). This chapter presents the application of the methodology to the flood forecasting model of the Klodzko catchment in Poland. The Klodzko catchment is located on the river Nysa Klodzka. The river serves as a highland tributary of the upper Odra River. The Klodzko catchment is a small mountain basin with a very short lag time between rainfall and runoff (see

Kundzewicz at al., 1999 and Szamalek, 2000). The river Nysa Klodzka and its tributaries are shown in Figure 5.1.

The area was heavily inundated by the great flood, which took place in the region in July 1997. The town of Klodzko (31,000 inhabitants) located on the river Nysa Klodzka was virtually ruined by this flood with several casualties and the destruction of numerous houses (Kundzewicz at al., 1999). Over 500 families in Klodzko lost virtually everything they owned. Since then flood forecasting and warning has been one of the important considerations for the people and authorities in the area.

The Klodzko model is a rainfall-runoff-routing type of model in which rainfall is a major time series input. The objective of the application is to assess the uncertainty in forecasted discharges due to the uncertainty in the forecasted precipitation. One important characteristic of a quick-response basin like the Klodzko is that the future situation of the flood depends heavily on the precipitation yet to occur. The nature of the basin provides a justification for its selection for the application of the methodology.

5.1.1 Description of the model

The rainfall-runoff-routing type flood forecasting model was built with HEC-1 (USACE, 1998) and HEC-HMS (USACE, 2000 and 2001) produced by the Hydrologic Engineering Centre of US Army Corps of Engineers. In particular, HEC-HMS (its Calibration Module) was used for model calibration and HEC-1 was used for simulation. HEC-1 is one of the most frequently used rainfall-runoff models in the United States (Melching et al., 1991). This model is a semi-distributed conceptual model for catchment modelling. The model simulates the precipitation-runoff and routing processes. A typical HEC-1/HEC-HMS representation of catchment runoff processes is shown in Figure 5.2. To represent the different components of the runoff processes, HEC-1 and HEC-HMS use different models including:

1. Models that compute runoff volume.

2. Models of direct runoff (overland flow and interflow).

3. Models of base flow.

4. Models of channel flow (routing model).

The flood forecasting model presented here covers the basin up to the hydrological station at Bardo (Fig. 5.1) on the Nysa Klodzka. The basin is subdivided into 9 subbasins. The subbasin areas range from 64 km^2 to 280 km^2 with a total area of 1744 km^2. Some of the parameters of the model are estimated by calibration and some are estimated from available data and engineering judgement. The calibration is performed using the calibration module of HEC-HMS. In particular, the SCS Curve Number, time of concentration, storage coefficient and Muskingum K are calibrated.

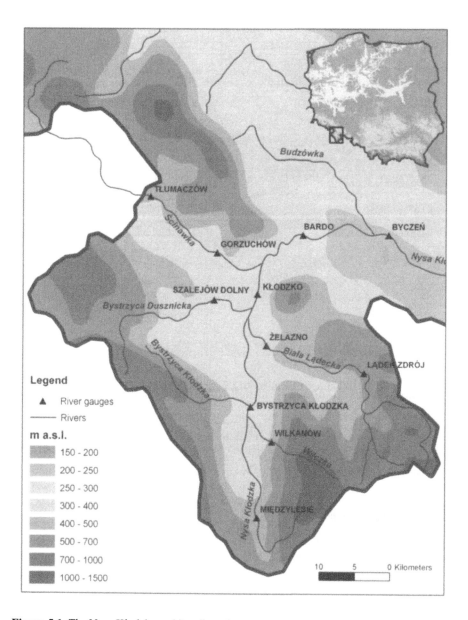

Figure 5.1. The Nysa Klodzka and its tributaries.

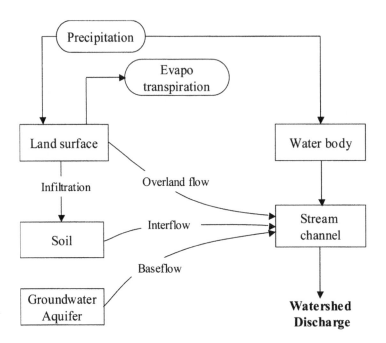

Figure 5.2. Typical representation of HEC-1 and HEC-HMS watershed runoff.

5.1.2 Methods used to model rainfall-runoff processes

Different options are available for selecting the methods for these models. The methods used in the present model are briefly described below. The selection of the methods in the present model is governed particularly by the availability of the data. The analysis of the influence of different methods is beyond the scope of this study.

Runoff volume computation model

The Soil Conservation Service (SCS) Curve Number (CN) method is used for the direct runoff volume (precipitation excess) computation, using the following equation:

$$P_e = \frac{(P - I_a)^2}{P - I_a + S} \tag{5.1}$$

where P_e is the accumulated precipitation excess (cumulative direct runoff volume in mm) at time t; P is the accumulated rainfall depth at time t; I_a is the initial abstraction (initial loss); and S is the potential maximum retention, that is the measure of the ability of a watershed to abstract and retain storm precipitation. The SCS developed the following empirical relationship between I_a and S

$$I_a = 0.2S \tag{5.2}$$

The maximum retention, S, is a function of the catchment characteristics. An intermediate parameter called curve number (CN) is used to compute S (in SI unit) as:

$$S = \frac{25400 - 254\mathrm{CN}}{\mathrm{CN}} \tag{5.3}$$

The CN values range from 100 for water bodies to approximately 30 for permeable soils with high infiltration rates.

Direct-runoff model

Clark's unit hydrograph (UH) method is used for the transformation of excess precipitation (runoff volume) to runoff. This method explicitly represents two critical processes of transformation: *translation* and *attenuation*. The former is the movement of the excess precipitation from its origin throughout the drainage to the catchment outlet, and the latter is the reduction in the magnitude of the discharge as the excess water is stored throughout the catchment. The transformation is defined in the model by two parameters: time of concentration, t_c, and the storage coefficient, R. The basin storage coefficient, R, is an index of the temporary storage of precipitation excess in the catchment as it drains to the outlet. Both of these parameters are normally estimated via calibration.

Baseflow model

The baseflow is computed using the exponential recession method (Chow et al. 1988). It defines the relationship of Q_t, the baseflow at any time t, to the initial baseflow (at time zero), Q_0, as:

$$Q_t = Q_0\theta^t \tag{5.4}$$

where θ is an exponential decay constant of the baseflow recession. In HEC-HMS, θ is defined as the ratio of the baseflow at time t to the baseflow one day earlier. In HEC-1 the θ is represented by the parameter called RTIOR, which is defined as the ratio of the current recession flow to the flow one hour later. The following equation can be used to convert an HEC-1 recession constant for use in HEC-HMS:

$$\theta = \frac{1}{(\mathrm{RTIOR})^{24}} \tag{5.5}$$

Other parameters required for baseflow computation for both HEC-1 and HEC-HMS are initial flow and threshold flow.

Routing model

The Muskingum method is used for the flow routing. This method is a commonly used hydrological routing method for handling a variable discharge-storage relationship (Chow at al., 1988). It uses a simple finite difference approximation of the continuity equation:

$$\frac{I_{t-\Delta t}+I_t}{2}-\frac{O_{t-\Delta t}+O_t}{2}=\frac{S_t-S_{t-\Delta t}}{\Delta t} \tag{5.6}$$

where $I_{t-\Delta t}$ and I_t are the inflow hydrograph ordinates at times t-Δt and t, respectively; and $O_{t-\Delta t}$ and O_t are the outflow hydrograph ordinates at times t-Δt and t, respectively. Similarly $S_{t-\Delta t}$ and S_t are the storage in the reach at times t-Δt and t, respectively. The storage is modelled as the sum of prism and wedge storage. The prism storage is the volume defined by the water surface profile at steady stage, while the wedge storage is the additional volume under the profile of the flood wave. The storage at time t, S_t, is defined as:

$$S_t = K[XI_t + (1-X)O_t] \tag{5.7}$$

where K is the travel time of the flood wave through routing reach and X is a dimensionless weight ($0 \le X \le 0.5$).

From Equations (5.6) and (5.7), it follows that

$$O_t = \left(\frac{\Delta t - 2KX}{2K(1-X)+\Delta t}\right)I_t + \left(\frac{\Delta t + 2KX}{2K(1-X)+\Delta t}\right)I_{t-\Delta t} + \left(\frac{2K(1-X)-\Delta t}{2K(1-X)+\Delta t}\right)O_{t-\Delta t} \tag{5.8}$$

Given inflow and outflow hydrographs, the Muskingum parameter K can be estimated. Once K is estimated, X can be determined by trial-and-error. These parameters can also be determined by calibration.

5.1.3 Uncertainty due to precipitation

Forecasting a flood using rainfall-runoff type models requires the forecasted precipitation over the forecast period. Over the last decade significant progress has been made in the quantitative forecast of precipitation using sophisticated radar technology. The uncertainty due to precipitation, however, remains a major part of the input uncertainty in such models. The impact of the uncertainty in precipitation from radar data on hydrological models is discussed by Borga (2002), Carpenter et al. (2001), Cluckie and Collier (1991) and Moore (2002), among others. The impact of the forecasted precipitation on the forecasted flood is particularly influential in quick-response basins.

The uncertainty in the forecasted precipitation results from the uncertainty in (1) the quantity, (2) the temporal distribution over the forecast period, and (3) the spatial distribution over the catchment. When a forecast is based on a probabilistic quantitative precipitation forecast, sampling methods such as the Monte Carlo technique are commonly used for the propagation of uncertainty through a forecasting model. In the availability of probabilistic forecasts of precipitations, the Bayesian theory based approach recently reported by Kelly and Krzysztofowicz (2000) and Krzysztofowicz (1999) may also be used. However it is not always possible to provide a reliable probabilistic assessment of the uncertainty in the precipitation, in which case alternative methods must be used.

The methodology presented in Chapter 4 of this thesis can be used in the framework of Monte Carlo (MC) simulation as well as the fuzzy Extension Principle (EP). The application of the methodology in both MC and EP frameworks is presented by Maskey and Price (2004). This method explicitly takes into account the uncertainty due to the unknown temporal structure of the precipitation. It is particularly important when the frequency of the available precipitation measurement is not sufficiently small. This methodology is independent of the structure of the forecasting model. That is, the methodology can be used with any rainfall-runoff-routing type deterministic model.

5.2 Implementation of the methodology for uncertainty assessment due to precipitation

This section presents the implementation of the methodology of uncertainty assessment presented in Section 4.1 to propagate the uncertainty due to the precipitation in the flood forecasting model (rainfall-runoff type) of the Klodzko catchment. Subsection 5.2.1 describes the reconstruction of precipitation by disaggregation and Subsection 5.2.2 describes the construction of a MF to represent the uncertainty in the magnitude of the precipitation. The propagation of the reconstructed precipitation using the fuzzy Extension Principle by the α-cut method is presented in Subsections 5.2.3, and the simplification applied to the methodology for the present application is presented in Subsection 5.2.4.

5.2.1 Precipitation time series reconstruction using temporal disaggregation for uncertainty assessment

Uncertainty in flood forecasting related to the uncertainty in time series inputs like precipitation is discussed in Section 4.1. The principle of the disaggregation of time series inputs for the treatment of uncertainty is also outlined in Subsection 4.1.1. The idea here is to divide the given temporal period into a fixed number of subperiods and to randomly disaggregate the given accumulated sum into as many subperiods, which aggregate up to the given accumulated sum. The disaggregated precipitations distributed over the subperiods (as a reconstructed time series) are then used in the rainfall-runoff model as inputs. Let an accumulated sum of the precipitation for a subbasin i ($i = 1, ..., m$) and for any time period be P_i. Denoting the disaggregated precipitation for a subperiod j by $p_{i,j}$ ($j = 1,..., n$; n = number of subperiods) we obtain

$$P_i = \sum_{j=1}^{n} p_{i,j} \qquad \forall i \qquad\qquad (5.9)$$

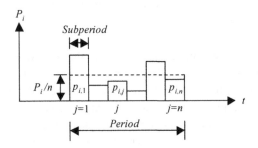

Figure 5.3. Average and disaggregated precipitations.

For simplicity coefficients $b_{i,j}$ are introduced, such that

$$p_{i,j} = P_i b_{i,j} \qquad\qquad (5.10)$$

where

$$\left.\begin{array}{ll} 0 \le b_{i,j} \le 1 & \forall(i,j) \\ \sum_{j=1}^{n} b_{i,j} = 1 & \forall i \end{array}\right\} \qquad\qquad (5.11)$$

The coefficients $b_{i,j}$ are generated randomly between zero and one. The Equations (5.9) and (5.10) allow the quantity $p_{i,j}$ to take any value from 0 to P_i. Whereas varying the coefficient $b_{i,j}$ over subperiods allows the possibility of different temporal distributions, varying the coefficients over subbasins allows for different spatial distributions. Thus by generating different values of $b_{i,j}$, as many temporal patterns as are needed can be generated.

5.2.2 Precipitation uncertainty represented by a membership function

Let P_i be the accumulated precipitation forecasted for a subbasin i ($i = 1,..., m$; m is the number of subbasins) for a given period T. In the absence of enough information to represent uncertainty in P_i probabilistically, it is assumed that it bears some error represented by dimensionless quantities e_1 and e_2 such that $0 \le e_1 \le 1$ and $e_2 \ge 0$. It is also assumed that the given value of P_i is its most credible (or most likely) value represented by $P_{i,\text{mc}}$. Then two other values minimum, $P_{i,\text{min}}$, and maximum, $P_{i,\text{max}}$, are defined as follows:

$$\left.\begin{aligned}
P_{i,\text{min}} &= (1-e_1)P_{i,\text{mc}} \\
P_{i,\text{max}} &= (1+e_2)P_{i,\text{mc}}
\end{aligned}\right\} \tag{5.12}$$

With these 3 values x_{mc}, x_{min} and x_{max} a triangular MF (Fig. 5.4) is assumed given by

$$\mu(P_i) = \begin{cases}
0 & \text{if } P_i < P_{i,\text{min}} \\[2mm]
\dfrac{P_i - P_{i,\text{min}}}{P_{i,\text{mc}} - P_{i,\text{min}}} & \text{if } P_{i,\text{min}} \leq P_i \leq P_{i,\text{mc}} \\[2mm]
\dfrac{P_{i,\text{max}} - P_i}{P_{i,\text{max}} - P_{i,\text{mc}}} & \text{if } P_{i,\text{mc}} \leq P_i \leq P_{i,\text{max}} \\[2mm]
0 & \text{if } P_i > P_{i,\text{max}}
\end{cases} \tag{5.13}$$

The qualitative meaning of the triangular membership function is the following. The "true" value of precipitation, P_i, is certainly included between $P_{i,\text{min}}$ and $P_{i,\text{max}}$ and is likely to be close to $P_{i,\text{mc}}$. The key words are "included" and "close". These words constitute the only information that one has about the problem (Revelli and Ridolfi, 2002). By doing so the uncertainty or imprecision associated with the single value of the forecast precipitation is represented using a fuzzy number on the basis of the minimum information available about the anticipated precipitation.

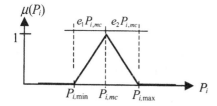

Figure 5.4. A triangular membership function with three values: $P_{i,\text{min}}$, $P_{i,\text{mc}}$ and $P_{i,\text{max}}$.

The triangular membership function is the simplest and frequently used form of the membership functions in many applications (Pedrycz, 1994). One obvious reason in using a triangular MF is due to the lack of information to justify the use of other shapes of MFs. A notable characteristic of a triangular MF is that it has a well focussed value corresponding to the maximum membership. Therefore, if some evidence supports more dispersed values for the maximum membership, the trapezoidal or parabolic forms of MFs might be more appropriate.

5.2.3 Algorithm for the propagation of uncertainty

The precipitation represented by a membership function and reconstructed using disaggregation into subperiods is propagated using the Extension Principle of fuzzy set theory. The Extension Principle is performed by the α-cut method. An example of an α-cut for a membership function and corresponding lower and upper bounds is shown in Figure 5.5.

Let a function f represents the rainfall-runoff-routing model with the precipitation as an input and runoff, Q, as an output:

$$Q = f(p_{i,j}; i = 1,...,m; j = 1,...,n)$$
$$= f[(p_{1,1},...,p_{1,n}),...,(p_{m,1},...,p_{m,n})]$$

$$(5.14)$$

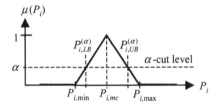

Figure 5.5. An α-cut level and corresponding upper and lower bounds. Definitions of $P_{i,\text{min}}$, $P_{i,\text{mc}}$ and $P_{i,\text{max}}$ are same as in Figure 5.4.

A general methodology for the propagation of the three forms of precipitation uncertainty through the model (Equation (5.14)) is presented in Subsection 4.1.4 for the EP-based approach. The methodology requires an algorithm for the determination of the maximum and minimum values of the model outputs (see Equation (4.16)). In this application a genetic algorithm (GA) is used for this purpose. Further description of the GAs used is presented in Section 5.3. The methodology (Subsection 4.1.4) adopted for the GA is outlined here:

1. Select a value of $\alpha \in [0,1]$ (called an α-cut level) for the MF of the input precipitation (Fig. 5.5) of all subbasins.

2. For the selected α-cut, determine for all $i = 1, ..., m$ the lower bound, $P_{i,\text{LB}}^{(\alpha)}$, and the upper bound, $P_{i,\text{UB}}^{(\alpha)}$ (See Fig. 5.5).

3. Start the GA with following parameters (decision variables) and constraints:

   Parameters for the input precipitation: $P_1^{(\alpha)},...,P_m^{(\alpha)}$ where $P_i^{(\alpha)} = [P_{i,\text{LB}}^{(\alpha)}, P_{i,\text{UB}}^{(\alpha)}]$ for all $i = 1, ..., m$.

Parameters for the disaggregation coefficients: $(b_{1,1},...,b_{1,k}),...,(b_{m,1},...,b_{m,k})$ where $b_{i,j} = [0, 1]$ for all $i = 1, ..., m$; $j = 1, ..., k$. The value of k depends on the type of the methods used for the generation of disaggregation coefficients. Three different methods for the generation of the disaggregation coefficients are presented in Subsection 4.1.6. For methods 1 and 2, $k = n - 1$, and for method 3, $k = n$.

4. The GA generates initial population (sets of parameter values).

5. Call an external program to adjust the disaggregation coefficients depending on the type of the method for determining the coefficients.

6. The GA evaluates the initial population using the forecasting model as an external program and a specified objective function (also called the fitness function in GA terminology). The fitness functions used are

$$f_t = \begin{cases} Q & \text{for maximum or upper bound} \\ -Q & \text{for minimum or lower bound} \end{cases} \tag{5.15}$$

where Q is the forecast discharge (model output) given by Equation (5.14).

7. The GA continues with crossover, mutation etc. (Subsection 5.3.2), generation of new population and re-evaluation of the fitness function until the termination criteria are met.

The output of the procedures 1 to 7 is either $Q_{min}^{(\alpha)} = Q_{LB}^{(\alpha)}$ or $Q_{max}^{(\alpha)} = Q_{UB}^{(\alpha)}$ depending on the type of the fitness function used. The whole procedure needs to be repeated for as many α-cuts as are needed to produce a complete MF for the output. See Figure 5.8 for a flowchart of this methodology combined with a conventional or normal GA.

In this methodology the number of parameters for the GA is given by

$$\begin{aligned} N_p &= m + (n-1)m \\ &= nm \end{aligned} \tag{5.16}$$

where N_p is the number of parameters for the GA, m is the number of subbasins and n is the number of subperiods.

5.2.4 Simplification of the methodology

An obvious drawback of the above methodology is that as the number of subbasins and subperiods increases, the search space becomes very large, which significantly reduces the performance of the genetic algorithm or any other algorithms used for the determination of the maximum and minimum. In an attempt to reduce the search space it was checked in several cases and it was found that the minimum and maximum of the model outputs (Q_{min} and Q_{max}) actually correspond to the lower bound ($P_{i,LB}^{(\alpha)}$) and the upper bound ($P_{i,UB}^{(\alpha)}$) values (for the given α-cut), respectively, of the precipitation determined in Step 2 of the algorithm. Therefore, instead of looking for all possible values of precipitations between LB and UB, $P_{i,LB}^{(\alpha)}$ is used to evaluate Q_{min} and $P_{i,UB}^{(\alpha)}$ is used to evaluate Q_{max}. This means that the model function is assumed to be monotonic with respect to the quantity of the accumulated precipitation of the forecast period. This simplification reduces the number of parameters for the GA (N_p) from nm to $(n-1)m$. This is viewed as an important issue because it helps to reduce the computational effort especially working with a big catchment consisting of many subbasins.

5.3 Genetic algorithms for minimum and maximum determination

Due to the widespread nature of the optimization problem for determining the minimum (or maximum) of a function, this has been one of the major fields of operational and mathematical research for decades. Algorithms for solving optimization problems range from linear, nonlinear to global. The suitability of the applications of these algorithms depends on the nature of the problem among other things. The global optimization algorithms (GOAs) have a particular advantage in solving problems in which other optimization techniques have difficulties when there exist multiple extrema and/or difficulties in defining functions analytically. Since GO algorithms do not require computation of derivatives, they can be a good alternative in solving optimization problems using off-the-shelf (black-box) software, where the details of the underlying algorithms may not be known. Successful applications of various GOAs are reported elsewhere, including Maskey et al. (2000a, b & 2002b). One of the most famous GOAs with widespread application is genetic algorithms (GAs). A comprehensive evaluation of optimization algorithms is not the intent of this study. However, the successful and extensive use of genetic algorithms (GAs) in various fields of engineering, including water related problems, inspired the author to choose a GA scheme as an optimizer.

5.3.1 Principles of genetic algorithms

Genetic algorithms, developed by Holland (1975), are search algorithms based on the mechanics of natural selection and natural genetics (Goldberg, 1989). A simple GA consists of

1. Random selection of the initial population,

2. Selection for mating,

3. Crossover, and

4. Mutation.

The initial population consists of sets of randomly selected solutions. A given solution set (set of parameters) is an individual in GA terms. An individual is represented by a chromosome set (also called strings) in GA operations. Chromosomes are made up of genes, like a solution set consists of parameter values. Thus, while parameters and solution sets are used in model terms, genes and chromosomes (also strings) are used in GA terms. Encoding and decoding are the processes to convert a solution set to strings and vice versa. Figure 5.6 illustrates the correspondence between model and GA terms.

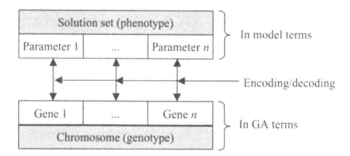

Figure 5.6. Model terms and GA terms for an individual of a population

The selection is a process in which parents (in pairs) are chosen from among individuals of the population for mating. The selection is made according to their objective function, popularly known as *fitness* function, values. The selected pairs (parents) from the mating pool are then crossed over (hence the process is called *crossover*) to produce new individuals (*offspring*), with a hope that the fit parents will produce even better fit children. Examples of one-point and two-point crossovers by bits-exchanges are shown in Figure 5.7. The offspring may also mutate. *Mutation* refers to the random distribution of genetic information. This is achieved by occasional (with small probability) random alternation of the value of a string position. In binary coding, this simply means changing a '1' to a '0' and vice versa.

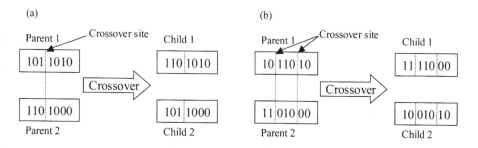

Figure 5.7. Examples of one-point (a) and two-points (b) crossovers. The crossover sites chosen are arbitrary.

The offspring obtained from selection – crossover – mutation operations replaces the unfit individuals (the solution sets with less fitness values) in the initial population and forms a new population for next the generation. A general structure of the genetic algorithms is as follows (Gen and Cheng, 2000):

```
Procedure: Genetic Algorithms
begin
  t ← 0;
  initialize P(t);
  evaluate P(t);
  while (not termination condition) do
  begin
    recombine P(t) to yield C(t);
    evaluate C(t);
    select P(t + 1) from P(t) and C(t);
    t ← t + 1;
  end
end
```

where $P(t)$ is the population for generation t, and $C(t)$ is the offspring produced in generation t.

5.3.2 GA versions used

Various versions of GA are used in practice. Comprehensive coverage of the topic can be found in Babovic and Keijzer (2000) and Goldberg (1989) among others. In the present experiment, two versions of GAs are used:

1. Conventional or normal GA (nGA).

2. Micro-GA (mGA).

The conventional GA is used with tournament selection, uniform crossover, two children per pair of parents, jump and creep mutations, elitism and niching. The tournament selection involves both random and deterministic selection features simultaneously. This method randomly chooses a set of chromosomes and picks up the best chromosomes for reproduction (Gen and Cheng, 2000). In uniform crossover, given two parent chromosomes of length 1, each parent copies 1/2 genes to each child, with the selection of the genes being chosen independently and randomly (Pawlowsky, 1995).

A combination of jump and creep mutation is used. In jump mutation the chromosome is chosen for mutation randomly. In creep mutation, on the other hand, the child will have a chromosome set which is different from one of its parents by only a small increment or decrement. As also applied by Carroll (1996) an equal probability was assigned for both jump and creep mutations. Elitism is used to guarantee that the chromosome set of the best parent (generated so far) is preserved. After the population is generated, the GA checks if the best parent has been replicated; if not, a random individual is chosen and the chromosome set of the best parent is mapped into that individual. The present GA also uses the operation called niching (sharing). Niching describes a process of identifying when individuals are converging on distinct optima and taking action to allow all the potential peaks to develop adequately. Goldberg and Richardson (1987) showed niching as an effective GA technique for multimodal problems. A sharing scheme by Goldberg and Richardson (1987) with a triangular sharing function was used.

The micro-GA (Krishnakumar, 1989) is used with tournament selection, uniform crossover, two children per pair of parents, elitism and niching. No mutation operation is used in this case. A micro-GA starts with a very small random population, which evolves in a normal GA fashion and converges in a few generations (typically 4 to 5). From this a new random population is created while keeping the best individual from the previously converged generation and restarting the evolution process (Yang et al., 1998).

A flowchart (Fig. 5.8) is presented here to illustrate the methodology presented in Subsection 5.2.3 combined with a conventional or normal GA. The Steps 1 to 7 in the flowchart are same as those in Subsection 5.2.3. In Step 2, the value of k is either n or $n - 1$, depending on the type of the method used for the generation of coefficients $b_{i,j}$ (see Subsection 4.1.6). Also in the figure, UB and LB represent upper and lower bounds respectively. In Step 9, the lower and upper bounds of the output discharges correspond to the minimum and maximum of the model outputs, respectively, which depends on the type of fitness function (Eq. (5.15)) used in Step 6.

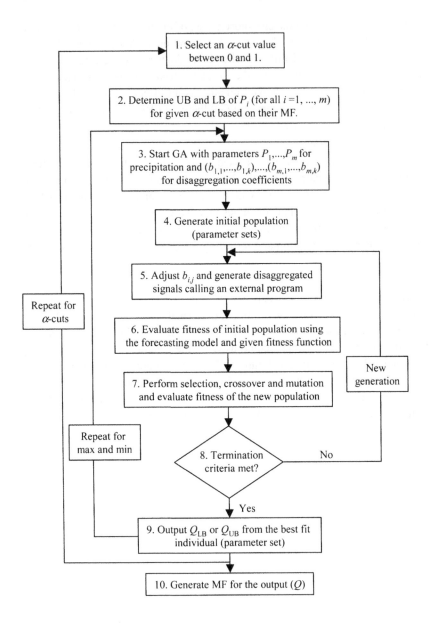

Figure 5.8. A flowchart for the implementation of the methodology in the framework of fuzzy Extension Principle supported by a normal GA for the propagation of precipitation uncertainty using disaggregation (see Subsection 5.2.3).

5.4 Application and results

The data corresponding to the great flood of July 1997 in the Nysa Klodzka River are used for the application. Observed precipitation data (cumulative for every three hours) are available. Due to the lack of forecasted precipitation data, the observed precipitation is assumed as a forecasted precipitation for the application of the present methodology. The MF representing the uncertainty in the quantity of the forecasted precipitation for each subasin is obtained by taking the error of +/-30 % in the given precipitation, i.e. $e_1 = e_2 = 0.3$ (Equation (5.12)). Although the error is taken arbitrarily, discussion with experts suggested that the assumption is not too far from reality. The starting date for simulation is 4 July 1997 at 06:00 hours, while the forecast of precipitation started on July 6 1997 at 00:00 hours (i.e. 42 hours from the point of start of simulation).

Since the interval of the available precipitation data is 3 hours, the size of the forecast period is also taken as 3 hours. The very short lag time between the rainfall and runoff justifies the use of the present methodology of uncertainty assessment with disaggregation. 3 subperiods of one hour each and 6 subperiods of half an hour each are used for disaggregation. It is to be noted that for each forecast, only the uncertainty in the forecast precipitation during the same forecast period is considered and no uncertainty is assumed in the precipitation previous to the current forecast period. For example, to forecast the flow for 3 hours, no uncertainty is assumed in all precipitation up to the time 0 hour; similarly to forecast the flow for 6 hours, no uncertainty is assumed in the precipitation up to the time 3 hours, and so on. A total of 20 forecasts (20 × 3 hours = 60 hours) from 6 July at 00:00 to 8 July at 09:00 have been carried out.

All the results presented and discussed here are based on the forecast at the hydrological station Bardo on Nysa Klodzka (Fig. 5.1). Firstly, the simulation is carried out to compute the discharge without considering any uncertainty in the precipitation. The simulated discharge together with the corresponding observed discharge is presented in Figure 5.9. The figure also shows the basin average precipitation on the negative y-axis. It is to be noticed that the simulation was carried out taking the accumulated precipitation in cach subbasin, and not with the basin average precipitation.

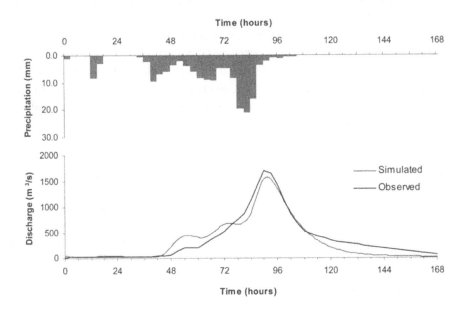

Figure 5.9. Observed and simulated discharges without considering uncertainty in the forecasted precipitation for the period 4 to 11 July 1997. The hour 0 refers to 4 July 1997 at 06:00 hours.

Secondly, the methodology is applied to estimate the uncertainty in the forecasted discharges due to the uncertainty in the forecasted precipitation. The results are characterised as "reconstructed" and "uniform". Whereas the "reconstructed" (i.e. with disaggregation) considers uncertainty due to the unknown temporal distribution of the precipitation, the "uniform" (i.e. without disaggregation) implies that average or uniformly distributed (throughout the subperiods) precipitation is used. The results produced by normal and micro GAs are very similar. Compared results for some cases are presented in Subsection 5.4.3. Since the results from both GAs are very close the results presented in Subsections 5.4.1 and 5.4.2 are all from the normal GA. The comparison of the results from the two versions of the GA is presented in Subsection 5.4.3.

5.4.1 Results with 3 subperiods

Figure 5.10(a) shows the upper and lower bounds of the forecasted discharges using reconstructed precipitation with 3 subperiods. Two cases with $\alpha = 0$ and $\alpha = 1$ are presented. In this graph, the lower bounds of the two cases are almost overlapping, whereas some deviations can be seen in the upper bounds.

Figure 5.11 is presented to further illustrate the differences in these two cases. It plots the uncertainty bounds in the forecasted discharges, i.e. $Q_{UB} - Q_{LB}$, for $\alpha = 0$ and $\alpha = 1$ against the forecast hours. The upper and lower bounds of the forecasted discharges with uniform precipitation are also presented in Fig. 5.10(b). For $\alpha = 0$ the upper and lower bounds show very little differences. For $\alpha = 1$ with the uniform precipitation, there exist no upper and lower bounds as there is only one value of precipitation, i.e. $P_{i,mc}$ (see Fig. 5.5).

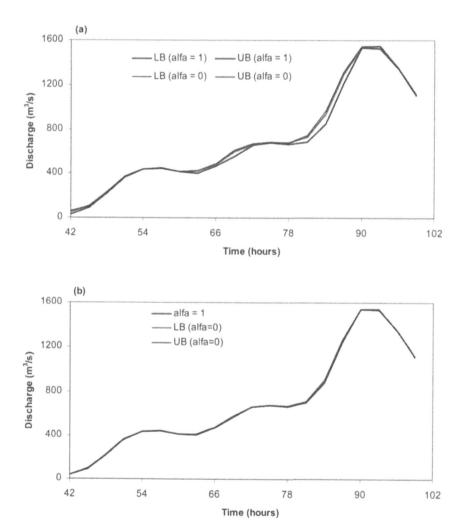

Figure 5.10. Upper bound and lower bounds of forecasted discharges for α (alfa)-cut levels 1 and 0: (a) with reconstructed precipitation, (b) with uniform precipitation. There are no upper and lower bounds for α-cut level 1 in the case of uniform precipitation.

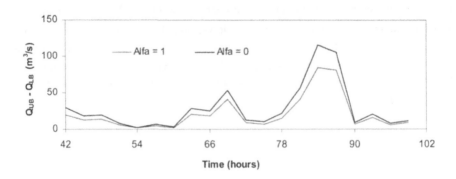

Figure 5.11. Uncertainty range in forecasted discharges (Q_{UB} - Q_{LB}) with reconstructed precipitation.

Figure 5.12(a) presents uncertainty bounds, $Q_{UB} - Q_{LB}$, for $\alpha = 0$ with reconstructed and uniform precipitations. The uncertainty bound for $\alpha = 1$ with reconstructed precipitation is presented in Figure 5.12(b). As there are no upper and lower bounds, there is no uncertainty bound with uniform precipitation for $\alpha = 1$. It is clearly seen that in all forecasts the output uncertainty is dominated by the case with a reconstructed precipitation. This suggests that the uncertainty due to the unknown temporal distribution can be more significant than the uncertainty in the quantity of the precipitation.

The results shown in Figures 5.10 through 5.12 are for α-cut levels 0 and 1 only. To construct a complete membership function (MF) of the output, the computation needs to be carried out at different level α-cuts. Three more α-cuts at $\alpha = 0.25, 0.5$, and 0.75 were selected. The MFs obtained for the forecasts at 81, 84 and 87 hours with reconstructed and uniform precipitations are shown in Figures 5.13(a) through (c). These figures further illustrate that the uncertainty due to the unknown temporal distribution is dominant over the uncertainty in the magnitude of precipitation.

Also of significance are the shapes of the MFs. The shapes of the MFs with uniform precipitation are fairly triangular showing that triangular MFs of inputs (precipitation) result in a triangular MF of output (discharge). However, with reconstructed precipitation the shapes are rather uniform, suggesting that more or less the same amount of uncertainty is added due to the uncertainty in the temporal distributions irrespective of the α-cut level.

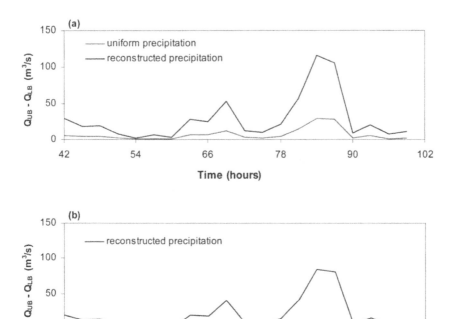

Figure 5.12. Uncertainty range in forecasted discharges (Q_{UB} - Q_{LB}): (a) for α-cut level 0, and (b) for α-cut level 1. There are no upper and lower bounds for α-cut level 1 in the case of uniform precipitation.

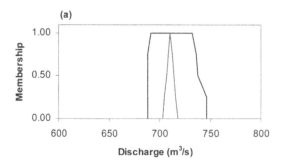

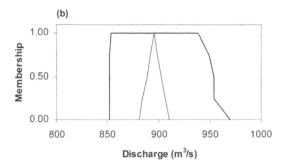

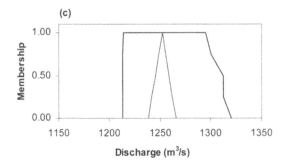

Figure 5.13. Membership functions of forecasted discharges with reconstructed precipitation (grey line) and with uniform precipitation (dark line): (a) forecast at 81 h, (b) forecast at 84 h, and (5) forecast at 87 h.

5.4.2 Results with 3 and 6 subperiods

A key issue in the application of this methodology is the selection of an appropriate number of subperiods. Increasing the number of subperiods increases the number of parameters for the search space and consequently the computational requirement. On the other hand, intuitively, increasing the number of subperiods may widen the uncertainty bounds in the output. To verify this, the results with 3 subperiods are compared with the results with 6 subperiods. With 6 subperiods, given 9 subbasins, the number of parameters increases from 18 (with 3 subperiods) to 45. To reduce the number of parameters it is assumed that the temporal pattern is the same for all subbasins. This assumption significantly reduces the number of parameters from 45 to 5 in the case of 6 subperiods and 18 to 2 in the case of 3 subperiods. It is however important to check the effect of this assumption. Therefore, the 3 subperiods case is repeated applying this assumption. The comparison of this result with the previous result for varying temporal patterns over subbasins is presented in Figure 5.14. The differences are minor except for forecasts at 84 and 87 hours. It is however important to notice that, in this particular example, the same temporal patterns over the subbasins resulted in larger uncertainty bounds than with varying patterns. The same temporal pattern is a special case of the varying temporal pattern. Therefore, the uncertainty in the case of a varying pattern should be larger than or at least equal to the case with same temporal pattern. This indicates a problem in the scheme used for finding the minima and maxima. For a GA scheme, given that other parameters are chosen appropriately, the solution of this problem normally requires more function evaluations. In this example, the maximum number of function evaluations is limited to 2000, which appeared not to be enough for the varying pattern.

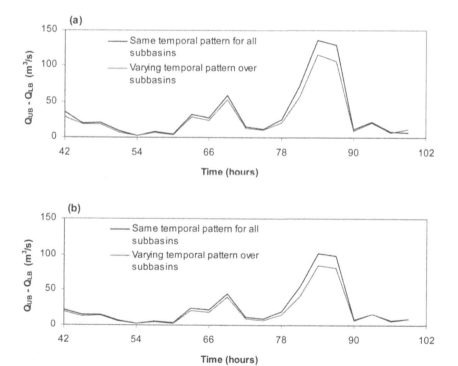

Figure 5.14. Uncertainty range in forecasted discharges (Q_{UB} - Q_{LB}) with reconstructed precipitation into 3 subperiods for: (a) α-cut level 0, and (b) α-cut level 1.

To illustrate this further, Fig. 5.15 is presented, which shows the graph of evaluated function values (minima and maxima) for different numbers of function evaluations. This example is for forecasts at 84 hours with 3 subperiods and varying temporal patterns over subbasins using the nGA. The figure clearly shows that the improvement in the function value is extremely slow after a certain number of function evaluations. Specifically, in the evaluation of the minimum function value (Fig. 5.15(a)) the gain in the function value is very slow after 1000 function evaluations. Similarly, in the evaluation of the maximum function value (Fig. 5.15(b)) the gain is very slow after 1200 function evaluations. In practical applications, it is normally not affordable to go for function evaluations as high as ten thousand as shown in Fig. 5.15. Moreover, the difference between the same temporal pattern for all subbasins and the temporal pattern varying over all subbasins (Fig. 5.14) is not very significant. Therefore, for the results presented in Fig. 5.14, the function evaluations are limited to 2000.

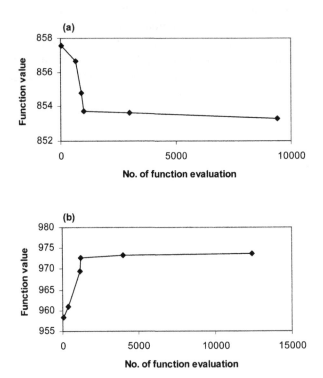

Figure 5.15. Performance of the algorithm for the determination of (a) minima, and (b) maxima, for the forecast at 84 h with 3 subperiods and varying temporal patterns over subbasins using the nGA.

Figure 5.16 shows the comparison of uncertainty bounds of cases with 3 and 6 subperiods for $\alpha = 0$ and 1. It can be observed that the increase in the uncertainty bound is not very significant compared to the increase in the number of subperiods.

This allows the limitation to a reasonably small number of subperiods without too much underestimation of the uncertainty in the output.

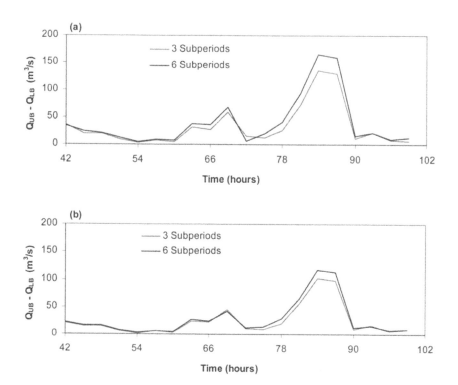

Figure 5.16. Uncertainty range in forecasted discharges (Q_{UB} - Q_{LB}) with reconstructed precipitation for: (a) α-cut level 0, and (b) α-cut level 1.

5.4.3 Results by normal GA and micro GA: a comparison

In general the results produced by the normal GA and the micro GA are very close to each other. The comparisons of the results given by the nGA and the mGA are shown in Figures 5.17 and 5.18 for three cases: (a) 3 subperiods with varying temporal distribution over subbasins, (b) 3 subperiods with same temporal distribution for all subbasins, and (c) 6 subperiods with same temporal distribution for all subbasins. Figure 5.17(a - c) compares the difference Q_{UB} - Q_{LB} produced by the nGA, the mGA and the best of the two GAs. Similarly, the difference in the lower bounds and in the upper bounds produced by the nGA and mGA are presented in Figure 5.18. The complete membership functions of the forecasted discharges given by the best of the two GAs are presented in Figure 5.19 (a – c). These MFs correspond to the case with 3 subperiods and with varying temporal distribution over subbasins.

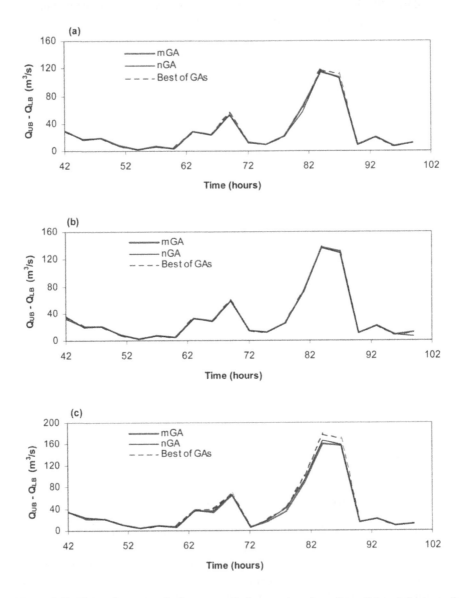

Figure 5.17. Uncertainty range in the output discharges given by mGA, nGA and the best of the two: (a) 3 subperiods, varying temporal distribution over subbasins, (b) 3 subperiods, same temporal distribution for all subbasins, and (c) 6 subperiods, same temporal distribution for all subbasins.

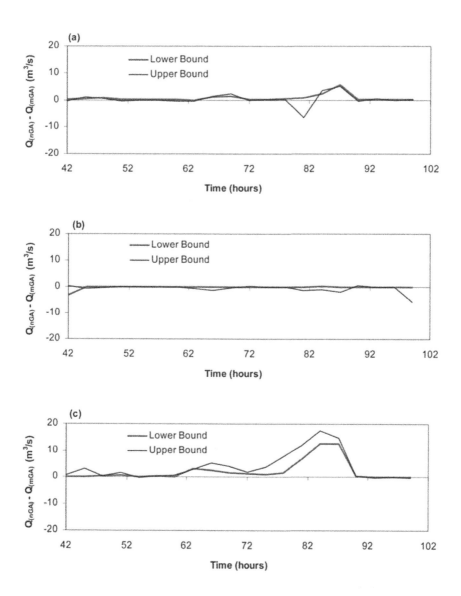

Figure 5.18. Differences in results from nGA and mGA on lower and upper bounds of the output discharges: (a) 3 subperiods, varying temporal distribution over subbasins, (b) 3 subperiods, same temporal distribution for all subbasins, and (c) 6 subperiods, same temporal distribution for all subbasins.

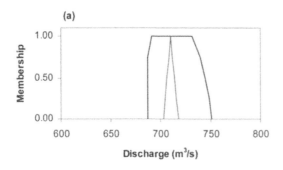

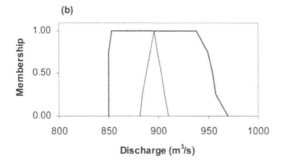

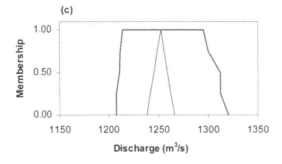

Figure 5.19. Best of the results of nGA and mGA: membership functions of forecasted discharges (3 subperiods and varying temporal distribution over subbasins) for (a) forecast at 81 h, (b) forecast at 84 h, and (5) forecast at 87 h. Dark likes for reconstructed precipitation and grey lines for uniform precipitation.

5.5 Conclusions and discussion

This chapter presented a methodology for the treatment of precipitation uncertainty in a rainfall-runoff-routing model in the framework of fuzzy set theory (using the Extension Principle) assisted by genetic algorithms. The methodology uses a reconstructed precipitation time series based on the disaggregation of precipitation into subperiods. This methodology is particularly useful in the absence of a probabilistic quantitative precipitation forecast. The methodology is independent of the structure of the forecasting model. In other words, it can be used with any rainfall-runoff-routing type deterministic model.

The results show the good potential of the fuzzy Extension Principle combined with a genetic algorithm for the propagation of uncertainty. The results also show that the output uncertainty due to the uncertainty in the temporal and spatial distributions can be significantly dominant over the uncertainty due to the uncertainty in the magnitude of the precipitation. This suggests that using space- and time-averaged precipitation over the catchment may lead to erroneous forecasts. The estimated uncertainty in the output may seem small compared to the magnitude of the flood. This is due to the relatively short forecast period (3 hours). Obviously increasing the forecast period significantly increases the uncertainty in the forecasted precipitation and thereby increases the output uncertainty. Moreover, it is to be noted that the estimated uncertainty is only due to the uncertainty in the precipitation. It does not include the parameter and model uncertainty.

An attempt has also been made to answer the question concerning an appropriate number of subperiods. Whereas the estimated uncertainty with reconstructed and uniform precipitation differed significantly (Fig. 5.12), the results with 3 and 6 subperiods showed only a small difference (Fig. 5.16). Therefore, in this particular example 3 subperiods seem good enough. In general, the determination of the number of subperiods should be governed by the consideration that: (i) the uncertainty should not be underestimated or overestimated beyond a reasonable limit, (ii) the computational requirements should not be too large, (iii) the subperiods should be large enough for the disaggregated precipitations to remain realistic, and (iv) the length of the subperiod should be of the same order of magnitude as the correlation time of the precipitation signal.

Chapter 6

APPLICATION: FLOOD FORECASTING MODEL FOR LOIRE RIVER (FRANCE)

Summary of Chapter 6

This chapter presents the results of the uncertainty analyses carried out on the flood forecasting model of Loire River in France. Three methods are applied: FOSM, Improved FOSM and fuzzy set theory and an expert judgment-based qualitative method. In the absence of sufficient information to represent uncertainty in the model inputs and parameters by probability distributions, the FOSM method is used as an appropriate tool for uncertainty estimation in the forecasts. Approximate estimates of the coefficient of variation of the discharges produced by rating curves are also assessed comparing the confidence intervals of the forecast water levels with the observed water levels. Separate analyses are carried out for the rising and subsiding flows. Although the FOSM method is found successful in most of the cases, a problem was encountered at a station for a particular flow condition near extrema due to the linearised assumption of the method. The Improved FOSM method (detailed in Section 4.2) is used to solve this problem. Results of the improved method are compared with the results of FOSM method and MC method. The effect of the different shapes of the PDFs and the size of perturbation ratios are also analysed. The expert judgement-based qualitative method using fuzzy set theory (detailed in Subsection 3.2.2) is also applied. The qualitative approach can incorporate the sources of uncertainty, which cannot be incorporated in the quantitative framework like FOSM method. Independent assessments of four experts are used for the analysis. The description of the Loire model and the sources of uncertainty are presented in Section 6.1. The applications of the FOSM, Improved FOSM and the qualitative methods are presented in Sections 6.2, 6.3 and 6.4, respectively. The conclusions and discussion are presented in Section 6.5.

6.1 Loire River flood forecasting model

The Loire River is the longest river (about 1,080 km) in France, which drains an area of about 117,000 square km. The Loire basin has a temperate maritime climate with heavy precipitation, including winter snowfall in the highlands that occupy its upper basin. The area of its headwaters is also subject to violent autumn storms from the Mediterranean. The river water level is usually highest in late winter, but there is no reliable rule; floods may occur in any month, though normally not in July and August

("Loire River." Encyclopaedia Britannica, 2001). The location of the Loire River is shown in Figure 6.1 and the tributaries of the river are shown in Figure 6.2.

The major historical floods of the Loire River in recent centuries occurred in 1846, 1856 and 1866, and minor ones but still with catastrophic impact in 1980 and 1982. During the flood of 1866, some 30,000 people were flooded in the middle Loire River basin. The number would be as high as 300,000 if such a flood occurred in the present day (Blancher et al., 2003). The Loire River basin is one of three pilot sites adopted by an EC project Operational Solutions for the Management of Inundation Risks in the Information Society (OSIRIS) (see Subsection 1.3.4). Uncertainty analysis of the existing flood forecasting system of the Loire River has been given significant importance in the project. The project has produced two prototypes for the Loire basin: one for the provision of user-friendly information on hydrological situation and the other for the tailoring of forecast information for a local diagnosis and for decision support (Collotte et al., 2003). A separate module has also been produced for the assessment of uncertainty in the forecast from the existing flood forecasting model (Maskey and Price, 2003a). Finally, the project increased the awareness of the need for a flood forecasting and warning system and the importance of uncertainty considerations in flood forecasting and warning.

The physics of the model is described in Subsection 6.1.1, and the various sources of uncertainty in the forecasts from the model are discussed in Subsection 6.1.2. The results of the uncertainty analyses are presented in Sections 6.2 through 6.4.

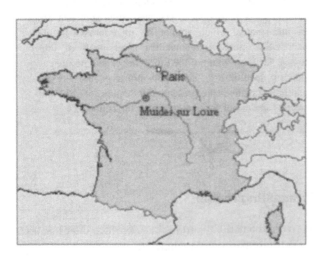

Figure 6.1. Location of Loire River (France).

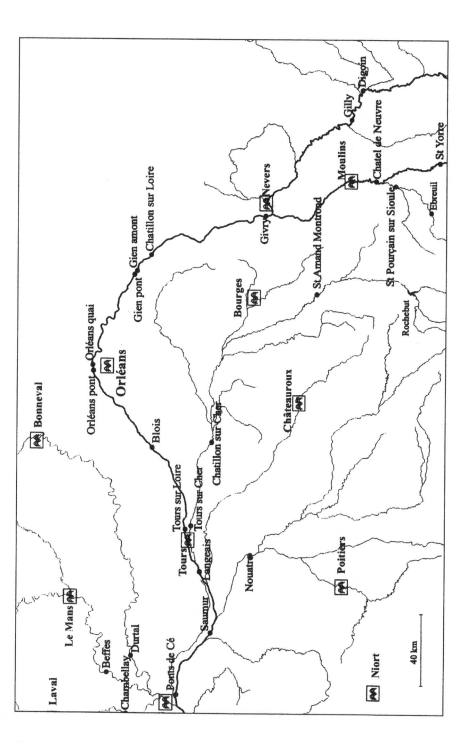

Figure 6.2. Loire River and its tributaries

6.1.1 Description of the model

The model under consideration forecasts water levels and discharges at a point using the water levels observed at certain points upstream of the river course. The basic steps involved in implementing the model are (i) the measurement of water levels at upstream stations, (ii) the estimation of discharges from the water levels, (iii) the propagation of flow to downstream stations, and (iv) the conversion of discharges to forecast water levels. The conversion from water level to discharge and *vice versa* is carried out using pre-established stage-discharge relationships (rating curves). A schematic diagram showing the processes involved in the model is presented in Figure 6.3.

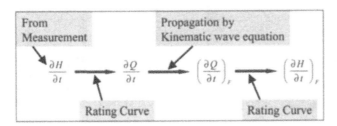

Figure 6.3. Schematic representation of the Loire River flood forecasting model. The subscript *F* represents forecasts.

The wave propagation step is based on the kinematic wave approximation, that is, the discharge is assumed to propagate downstream according to:

$$\frac{\partial Q}{\partial t} + c\frac{\partial Q}{\partial x} = 0 \qquad (6.1)$$

In characteristic form, this becomes:

$$\frac{dQ}{dt} = 0 \ \text{ along } \ \frac{dx}{dt} = c \qquad (6.2)$$

which gives the possibility to forecast the discharge at a given point (x, t) using the measurements at another point located upstream at a distance D:

$$Q(x,t) = Q(x - D, t - D/c) \qquad (6.3)$$

It is therefore sufficient to measure the discharge at a given distance upstream and to propagate it downstream with a lag time $D/c = T$.

However, the equation is not used in the form above, because a previous wrong forecast can never be updated with the current measurement of the discharge. What is used instead is the time variation of the discharge. Equation (6.1) is differentiated with respect to time. Assuming that c remains constant in time, this yields:

$$\frac{\partial}{\partial t}\left(\frac{\partial Q}{\partial t}\right) + c\frac{\partial}{\partial x}\left(\frac{\partial Q}{\partial t}\right) = 0 \tag{6.4}$$

This allows the forecast to be corrected for a previous erroneous forecast by incorporating the current value of the discharge. Indeed, Equation (6.4) can be discretized as:

$$Q(x, t + \Delta t) = Q(x, t) + Q(x - D, t - T + \Delta t) - Q(x - D, t - T) \tag{6.5}$$

A temporal interpolation is needed if the time instant $t - T + \Delta t$ does not correspond to a time where the discharge has been measured. The temporal interpolation of Equation (6.5) can be carried out as follows:

$$Q(x, t + \Delta t) = Q_1 + (1 - r)Q_2 + (2r - 1)Q_3 - rQ_4 \tag{6.6}$$

where $Q_1, ..., Q_4$ are defined as follows:

$$\left.\begin{array}{l} Q_1 = Q(x, t) \\ Q_2 = Q(x - D, t - (n - 1)\Delta t) \\ Q_3 = Q(x - D, t - n\Delta t) \\ Q_4 = Q(x - D, t - (n + 1)\Delta t) \end{array}\right\} \tag{6.7}$$

Here, the integer value n and the residual fraction r are defined as

$$n = \text{int}\left(\frac{T}{\Delta t}\right) \tag{6.8}$$

$$r = \frac{T}{\Delta t} - n \tag{6.9}$$

6.1.2 Sources of uncertainty

The various sources that lead to the uncertainty in the forecasts from the model discussed above are identified as follows:

1. Water level measurements
2. Rating curve conversions
3. Propagation model

4. Other sources due to the ignored processes in the conceptualisation of the model.

The observed water levels at various time and locations are the only inputs to this model. Therefore, the uncertainty due to the imprecise measurement of water levels is the input uncertainty for this model. The uncertainty due to the rating curve conversions and the process of propagation are considered as parameter uncertainty. The uncertainty due to the ignored processes in the conceptualisation of the model is referred to here as other sources. They are the ignored effects such as the interaction with confluences, the attenuation of the flood wave and so forth.

Contrary to the rainfall-runoff type model presented in Chapter 5, in the present model the uncertainty due to the uncertainty in the inputs is not very significant. This is due to the facts that, in this particular model, the inputs are all measured (not forecasted) and that the water level measurements are reasonably precise. According to the information provided by the DIREN Centre in Orleans (France), the accuracy of the gauges for water level measurement is ± 1 cm. Also the calibration of the gauges has been done recently and properly. The sources of uncertainty in the propagation of the flow and in the stage-discharge conversions are the major sources of uncertainty for this model.

Pre-established rating curves, which are revised every two years, are used. The conversion from water level to discharge is done using rating curves constructed from regular sampling measurements. The uncertainty due to water depth to discharge conversion can be particularly significant for high water levels when the extrapolation of the curve is required.

The assumption of a constant celerity c (and consequently of a constant propagation time T) is obviously a major limitation of the model, because in reality the wave propagation speed is related to the flow rate (see, for example, Cunge et al., 1980). As acknowledged by the users of the model, the value of T can only be indicative, based on experience and typical values assessed in the past. The value of T particularly influences the arrival time of the flood, which consequently may give rise to an inaccurate flood estimate for each forecast. In flood crisis management it is also very important to know the time available (e.g. for evacuation or for any safety measures to be taken) before the flood level reaches a given level and the time the flood level will remain above the given threshold. Consequently, there was a strong need for an uncertainty analysis in T.

In this study the uncertainty in the forecasts from the model is analysed using the probability theory based FOSM method and the fuzzy set theory and expert judgement -based qualitative method. The IFOSM method detailed in Section 4.2 is also applied to this model.

For the application of the FOSM method only the parameter uncertainty (in the rating curve conversion and in the propagation) is considered. The input uncertainty in the water level measurements is ignored assuming it is insignificant, and the "other sources" of uncertainty are excluded as they are beyond the scope of the FOSM framework. In the application of the qualitative approach based on expert judgement, all four sources of uncertainty listed above are assessed. The applications of the FOSM and IFOSM methods are presented in Sections 6.2 and 6.3, respectively, and the application of the qualitative method is reported in Section 6.4.

6.2 Uncertainty analysis using the FOSM method

Methods of uncertainty estimation based on probability and fuzzy set theory are presented in Chapter 3. The FOSM method is one of the widely used probability theory-based method for uncertainty propagation through a model. The FOSM method propagates only the moments of the distribution, and therefore does not require information about the complete distribution of the parameters. This characteristic of the FOSM method makes it suitable to apply to the Loire model. The uncertainty in the forecasted water level from the Loire model due to the parameters of the flow propagation and rating curve conversions is analysed. The river reach between Givry and Orleans is considered, which is subdivided into 2 reaches: Givry to Gien and Gien to Orleans. The results of the uncertainty assessment presented here are for the water levels at the station Orleans.

6.2.1 Description of data

The flow propagation time T between any two consecutive stations is the parameter for the propagation model. Very limited information is available about the estimates of propagation time between any two stations and therefore the actual shape of the PDF of T is unknown. From the experience of the users of the model, three different values of T (minimum, most likely and maximum) are assessed corresponding, respectively, to typical low, medium and high discharges experienced in the past. From this information, a triangular PDF is assumed and the mean and standard deviation for each value of T are determined accordingly.

As mentioned above the pre-established rating curves are used to obtain discharges from water levels and *vice versa*. The only information available about the precision of the rating curves is that the calculated discharges could vary by 10% to as much as 30%. This information is insufficient to make a sensible estimate about the moments of the distributions for the results of the rating curve conversions. Therefore, instead of estimating directly the standard deviation or variance of the distribution, an estimate about the coefficient of variation (CoV) is used. The conventional definition of the CoV is used as follows:

$$\gamma_Q = \frac{\sigma_Q}{\overline{Q}} \qquad\qquad (6.10)$$

where γ_Q, σ_Q and $\overline{Q}$ are, respectively, the coefficient of variation, standard deviation and mean value of the discharge estimated using the rating curve.

There are five uncertain parameters. Three parameters on rating curve conversions, one each for the three stations, and two parameters on the flow propagation time for the two reaches. The estimated mean values and the standard deviations for the flow propagation time T for the two segments are given in Table 6.1.

The flood data of April 1998 are used. In the first part of the assessment, an assumption of 0.1 CoV is used to estimate the uncertainty in the forecast water levels. The 95 percent confidence bounds are computed from the estimated uncertainty and are plotted against the measured water levels at different forecast hours. The forecasts are carried out from 4 hours to up to 48 hours for two cases: (i) for the rising flow situation and (ii) for the subsiding flow situation.

Table 6.1: Properties of the propagation time T for different river reaches (means and standard deviations are computed assuming triangular PDFs).

River reaches	Properties of the propagation time, T (h):	
	Mean	Standard deviation
Givry–Gien	24	3.27
Gien–Orleans	20	2.45

6.2.2 Results of analysis

The uncertainty contributions of each of the five parameters are presented in terms of the standard deviation of the water levels in Figure 6.4. For the rising flow case, the contribution of the rating curve conversions remains more or less constant, whereas the contribution of the propagation time is increasing with increasing forecast lead-time. The trend however, is not very distinct for the subsiding flow case.

The total estimated uncertainty (standard deviations) together with the forecasted water levels are shown in Figure 6.5 for the cases with rising and subsiding flows. The uncertainty is clearly larger for the rising flow than for the subsiding flow. The figure also shows that the estimated uncertainty by and large is increasing with increasing forecast lead-time.

The 95% confidence bounds for the forecasts at different lead-times are plotted with the corresponding observed water levels, both for the rising and subsiding flow cases (Fig. 6.6). Triangular PDFs of the forecasted water levels are assumed to compute the confidence bounds. It can be clearly seen that the observed water levels are well within the 95% confidence limits in the case of the subsiding flow, whereas they are mostly outside the limits in the rising flow case. By limiting the analysis to the five uncertain parameters considered in this exercise and assuming that the estimate of uncertainty in the propagation time is reasonably correct, the uncertainty assumed in the rating curve conversion is not large enough to keep the observed water levels within 95% confidence limits in the rising flow example. This conclusion is in agreement with the assumption that a large uncertainty in the rating curve conversion can be expected for the higher water levels. This is due to the fact that as the data for the extreme events (very high or very low flow) are naturally sparser, the rating curves in these regions are produced either by extrapolation or interpolation with very limited data. Consequently, this leads to a larger uncertainty in the estimated flows.

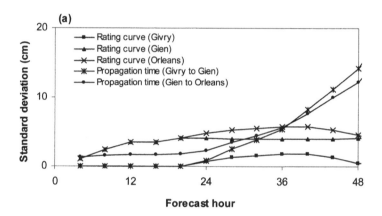

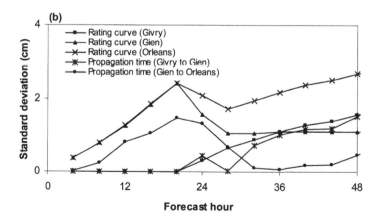

Figure 6.4. Uncertainty (standard deviation) in forecast water levels due to various uncertain parameters: (a) during rising flow, and (b) during subsiding flow situations.

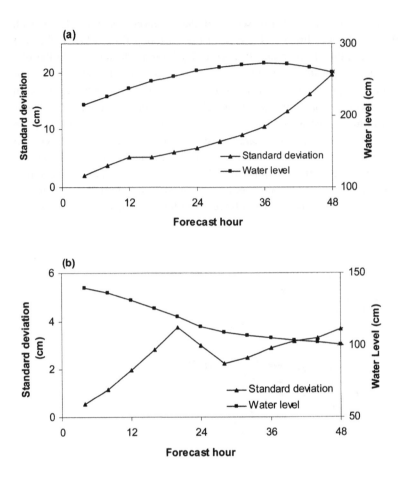

Figure 6.5. Forecasted water levels and total uncertainty (standard deviation) in the forecasted water levels due to uncertain parameters: (a) during rising flow, and (b) during subsiding flow situations.

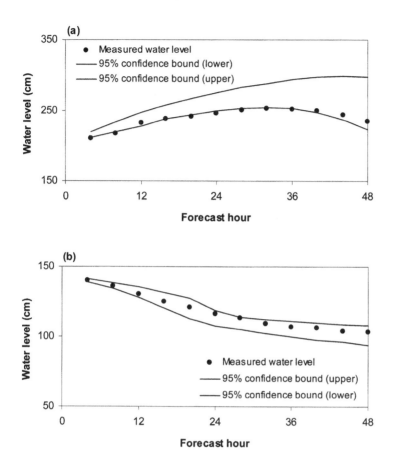

Figure 6.6. Measured water levels and 95% confidence bounds to the forecasted water levels: (a) during rising flow, and (b) during subsiding flow situations.

The second part of the analysis is focused on finding an appropriate value of the coefficient of variation for the rating curve conversions so as to keep the observed water levels within the 95% confidence limits for the rising flow case. The uncertainty contributions of the rating curve conversions to the uncertainty in the forecasts are computed considering different levels of uncertainty in the rating curves with the CoV ranging from 0.1 to 0.3. The results are shown in Figure 6.7. It can be observed that the increase in the uncertainty contribution is more or less linear with the increase in the CoV.

95% confidence bounds are computed for the total uncertainty with different levels of uncertainty in the rating curves. Figure 6.8 shows the observed water levels with the 95% bounds for the CoV values of 0.1, 0.15 and 0.2 in the rating curve conversions. It is seen that the observed water levels can be well within the 95% confidence bounds with the CoV of 0.2.

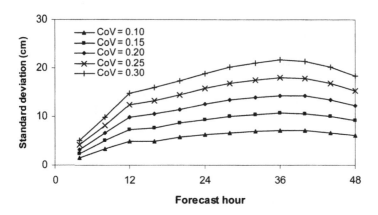

Figure 6.7. Uncertainty (standard deviation) of forecasted water levels due to uncertainty in rating curves (total of 3 rating curves at Givry, Gien and Orleans) for different coefficient of variations.

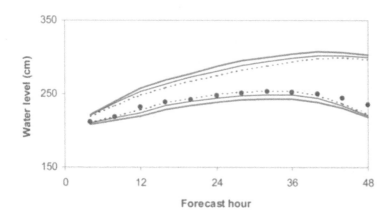

Figure 6.8. Measured water levels (points without lines) and 95% confidence bounds with various coefficients of variations 0.10 (discontinuous lines), 0.15 (continuous lines) and 0.20 (thick grey lines) in water levels or discharges computed using rating curves.

6.2.3 Conclusions

The present analysis demonstrated the useful application of the FOSM method for the estimation of uncertainty in the forecasts of a flood forecasting model for which detailed information about the uncertainty of the input parameters is not available. The analysis showed that the forecast uncertainty is higher for high flows than for low flows. It also showed that the uncertainty in the forecast is generally increasing with an increasing forecast horizon. These two conclusions are intuitive. Furthermore, it is also shown that the observed data and the confidence intervals of the estimated uncertainty can be used to make a reasonable estimate of the uncertainty in some of the inputs, when enough information is not available.

6.3 Application of the improved FOSM method

The application of the FOSM method to the Loire model presented in Section 6.2 shows that it can be a good tool for the assessment of uncertainty in flood forecasts in the absence of detailed information about the uncertainty in the inputs and parameters of the model. However, in some cases the estimated uncertainty may suffer from the limitations of the method. The merits and limitations of the FOSM method are discussed in Section 4.2. In particular the limitation of the FOSM method arises from the linearisation of the model function. One example of this was encountered when applying the method to the Loire model to assess uncertainty at the station Givry for a specific flow condition. In an effort to reduce such limitations the IFOSM method is developed and detailed in Chapter 4. The IFOSM method is applied as an alternative to the FOSM method. Since the MC method can provide values close to the analytical solution, it was used as a standard method against which results from the FOSM method and the IFOSM method can be compared.

6.3.1 Description of data

The data of April 1998 flood are used. The uncertainty in the propagation time T for seven reaches of the Loire River upstream of Givry are considered. Like the previous example, the only information available about the estimates of the Ts are the minimum, maximum and the most likely values corresponding respectively to typical low, medium and high discharges experienced in the past. Therefore, the means and the standard deviations are derived assuming a triangular PDF based on the minimum, maximum and most likely values. The estimated properties of the propagation time for various reaches of the river are presented in Table 6.2.

Table 6.2: Properties of the propagation time T for different river reaches (means and standard deviations are computed assuming triangular PDFs).

River reaches	Properties of the propagation time, T (h):	
	Mean	Standard deviation
Villerest–Digoin	14.0	2.04
Digoin–Gilly	4.0	0.61
Gilly–Givry	24.0	3.27
Ebreuil–St. Pourcain	12.0	1.63
St. Pourcain–Chatel	7.0	1.02
St. Yorre–Chatel	5.0	0.82
Chatel–Givry	12.0	1.63

6.3.2 Results of the analysis

The standard deviations of the forecasted water levels for different forecast hours are estimated using the IFOSM method. The results are then compared to the estimated uncertainty from the FOSM and MC methods. The influence of the shapes of the PDF (uniform, triangular and normal) and the sensitivity of the estimated uncertainty with respect to the perturbation ratio are also analysed.

Results comparison: FOSM, Improved FOSM and MC methods

The results from the FOSM, IFOSM and Monte Carlo methods are compared in Figure 6.9(a, b). Figure 6.9(a) displays the results obtained using uniform PDFs and Figure 6.9(b) presents the results obtained using triangular PDFs. In most cases the results from the FOSM and MC methods are significantly different. In particular, the FOSM method fails to provide a proper estimate of the uncertainty for forecast horizons of 4 and 16 h. This is due to the presence of a local maximum in the forecast water levels. In particular, the FOSM method underestimates the uncertainty because the magnitude of the derivative is underestimated. The problem is eliminated by the application of the IFOSM method. The results from the IFOSM method are very close to those of the MC method when the uniform PDF is used (Fig. 6.9(a)). The results of the improved method with the triangular PDF (Fig. 6.9(b)) are also satisfactory, but they tend to overestimate the uncertainty compared with the uniform PDF. This could be expected because the support of the triangular PDF is wider than that of a uniform PDF for a given standard deviation. Still, the results are better than the results from the FOSM method.

Effect of the shape of the PDF

As mentioned earlier, the FOSM method does not use any particular assumption about the PDF of the input variables. Consequently, only the first- and second-order moments of the output are known. The influence of an *a priori* assumption about the nature of the input PDFs on the computed uncertainty in the output is assessed. For a given mean and standard deviation, three different PDF shapes were assumed for the travel time T: uniform, triangular and normal. For each of these three assumptions, the MC simulation provided the reference estimate of the uncertainty in the forecast water levels. The result is shown in Figure 6.10. The computed standard deviations of the forecast water levels (at various horizons) for all three PDFs are very close, showing that the shape of the distribution is not the most influential factor in this particular application. Therefore, for computational simplicity, a normal distribution can be replaced with a uniform or triangular one in the calculations without leading to a large error in the uncertainty analysis.

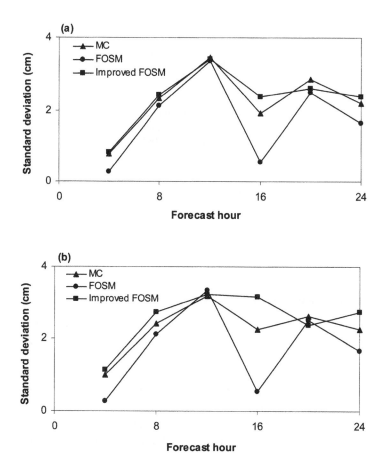

Figure 6.9. Standard deviations of forecast water levels given by the FOSM, IFOSM and Monte Carlo methods (a) with a uniform PDF; and (b) with a triangular PDF.

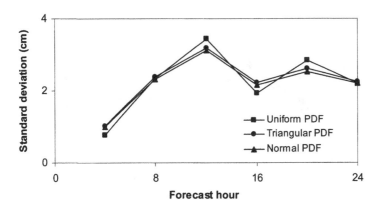

Figure 6.10. Standard deviations of forecast water levels given by Monte Carlo (MC) method for uniform, triangular and normal PDFs.

Sensitivity of the FOSM method to perturbation ratio

It is important to note that the results obtained from both the FOSM and IFOSM methods vary with the size of the perturbation (because the approximate function of the real function depends on the size of the perturbation). A measure of the perturbation is the perturbation ratio (PR), that is, the ratio of the actual perturbation to the standard deviation of the parameter. The values shown in Figures 6.9(a) and (b) are the best results on average for all forecast horizons, when taking the same PR for all forecast horizons. The results can be further improved if different PRs are used for different forecast horizons, but this yields an undesirable increase in computational cost.

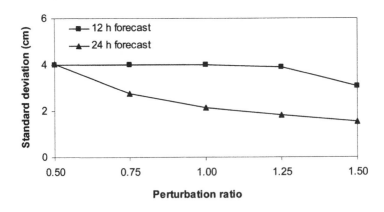

Figure 6.11. Standard deviation *vs* perturbation ratio for FOSM method for forecast water levels at 12 and 24 h.

Figures 6.11 and 6.12(a, b) show the variation of computed standard deviations on 12- and 24-h forecast water levels with respect to the PR applied for the FOSM and IFOSM methods, respectively. Clearly, the results are sensitive to the PR. In the present example, the result for the 24-h forecasts is more sensitive than that for the 12-h forecast. The standard deviation obtained from the MC method with a uniform PDF is 3.5 cm for the 12-h forecast and 2.2 cm for the 24-h forecast. When triangular PDFs are used, the computed standard deviation is 3.19 and 2.26 cm for 12- and 24-h forecasts respectively. As to which PR gives the best result, this may vary from case to case. In the present example, the FOSM method reaches the closest value to the MC results for a PR ranging from 1 to 1.5. The IFOSM method reaches the closest values for PR values ranging from 1.5 to 1.75.

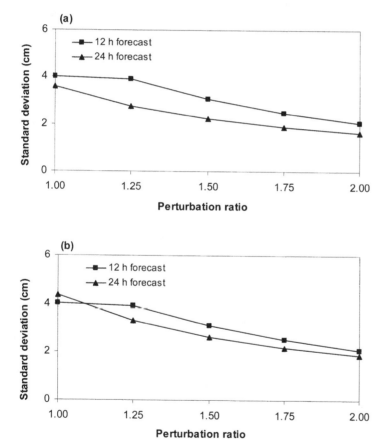

Figure 6.12. Standard deviation *vs* perturbation ratio for IFOSM method for forecast water levels at 12 and 24 h: (a) with uniform PDF, (b) with triangular PDF.

6.3.3 Conclusions

The IFOSM method produced a satisfactory result when solving a particular problem that is encountered in the application of the FOSM method to a flood forecasting model for the Loire River. The problem concerns the misleading estimation of uncertainty when the average value of the input variable corresponds to a maximum or minimum value or to regions where the slope of the function is very mild and varies gradually compared to the effects of a large curvature (non-linearity) of the function. There is no significant computational burden in implementing the IFOSM method because it requires only three function evaluations for each input (uncertain) variable, compared to two for the FOSM method. In the presented examples, the unknown PDFs are approximated by a uniform and a triangular PDFs. In principle the method can be applied to all possible PDFs. However, the more complex the PDF the more complicated the derivation of the solution. When the PDF is unknown, assuming a complex distribution cannot be justified more than choosing a simpler one. Obviously the method is more suitable for the problems that are less sensitive to the shape of the PDF, like the one presented in this example. Selecting an appropriate perturbation size may compensate, to some extent, for the error introduced by simplifying the PDFs.

Both the FOSM and IFOSM methods are sensitive to the size of the perturbation. Therefore, it is important that before actually applying these methods to practical problems an appropriate size of the perturbation must be selected. This problem has been discussed in the literature but the recommendations differ according to the nature of the cases. For example, Morgan and Henrion (1990) suggested a perturbation comparable to the range of variation; Haldar and Mahadevan (2000a) adopted a perturbation equal to the standard deviation of the variable, and Haldar and Mahadevan (2000b) suggested a perturbation equal to one tenth of the standard deviation in a particular problem of solid mechanics. Obviously, the best value of the perturbation varies from case to case. Several experiments carried out with this flood forecasting model show that the best value of the PR is in the range from 0.75 to 1.5 for the FOSM method and from 1.25 to 1.75 for the IFOSM method. Generally the IFOSM method gives better results for large perturbation ratios, because the approximate function becomes smoother when the three points used for the second degree reconstruction are taken further apart. Conversely, the slopes and curvature of the approximate parabolic function can be very large when the perturbation ratio is very small. Therefore, the IFOSM method is best-suited for long-term forecasts. It is however advisable to use some other standard method, for example the MC method, to fix beforehand the best value of the perturbation for a range of possible scenarios. This has particular advantages in the problem of real time application, including flood forecasting, where a decision has to be made in a very short time. The MC method requires a large number of model runs depending on the complexity of the model and the level of accuracy anticipated. The IFOSM method, if calibrated properly for the perturbation size, can generate results as good as the MC method for a greatly reduced computational effort.

6.4 Uncertainty analysis using qualitative method

Most physically-based models are complex by nature and involve many input variables and model parameters that cannot be determined precisely. Uncertainty in some of such parameters cannot be estimated quantitatively, such as using PDFs due to insufficient data. For a complex model, particularly in real time use, the MC-type methods may not be feasible due to time constraints. What is usually done in practice is to exclude some of the parameters/variables (that are supposed to have less influence on the uncertainty) from the analysis. Furthermore, there may exist some sources of uncertainty other than the model parameters and variables, which obviously cannot be incorporated in such a quantitative method. Krzystofowicz (1999) also concluded that the uncertainty assessed by the methods like the MC method alone is not sufficient unless the uncertain parameters that are ignored are insignificant.

In such situations the fuzzy set theory and expert judgement-based qualitative method presented in Subsection 3.2.2 can be useful. Given the qualitative assessments of the parameter uncertainty by experts, this type of qualitative method allows for the estimation of the uncertainty due to all recognisable sources without a significant increase in the computation time.

In this section the results of the application of a qualitative method based on fuzzy set and expert judgement applied to the estimation of the uncertainty in the forecasting of floods using the Loire model are reported. The forecasts at the station Orleans are considered, and the forecasts are assumed to be on a daily basis (24 hours forecast horizon). Subsection 6.4.1 presents the evaluations by experts on different parameters of uncertainty with respect to their contributions to the uncertainty in the output. The results are presented in Subsection 6.4.2.

6.4.1 Expert evaluation

Four parameters have been identified for the qualitative evaluation of uncertainty. They are (i) the imprecision in the measurement of water levels, (ii) the uncertainty due to rating curve conversions, (iii) the uncertainty due to propagation of flow, that is the propagation model, and (iv) the uncertainty due to other reasons. Each parameter has two or more sub-paramters (Table 6.3). The evaluation is carried out on *Quality* and *Importance* of the uncertainty parameters regarding their contributions to uncertainty in the forecast. Five linguistic variables are used for the qualitative evaluations of both quality and importance. The linguistic variables used for the evaluation of *Quality* are *Very Good, Good, Acceptable, Bad* and *Very Bad* and those used for *Importance* are *Very Large, Large, Moderate, Small* and *Very Small*. The fuzzy membership functions representing these variables are given in Figure 6.13. The *Importance* is evaluated for the parameters and the *Quality* is evaluated for the sub-parameters. The *Quality* of the parameter is derived from the *Qualities* of the sub-parameters using Equation (3.22). Four different individuals experienced in the forecasting system, here referred to as *experts*, took part in the evaluation. The

evaluation is presented in Table 6.3. In the table, the evaluation of *Importance* which
is for the paramter is given in italics.

Table 6.3: Experts' evaluation on *Quality* and *Importance* of uncertain parameters of the
flood forecasting model of Loire River (France).

Parameters and sub-parameters	Importance of parameters (in italic) and Quality of sub-parameters			
	Expert 1	Expert 2	Expert 3	Expert 4
Water level measurement	*Very Large*	*Very Large*	*Very Large*	*Very Large*
Main Loire	Very Good	Good	Good	Very Good
Tributaries	Very Good	Good	Acceptable	Good
Rating curve	*Large*	*Very Large*	*Very Large*	*Very Large*
Main river	Good	Good	Acceptable	Very Good
Tributaries	Acceptable	Good	Acceptable	Acceptable
Propagation model	*Large*	*Large*	*Large*	*Large*
Equation	Acceptable	Acceptable	Very Bad	Good
Assumptions	Acceptable	Bad	Bad	Acceptable
Parameters	Acceptable	Acceptable	Bad	Good
Discretisation	Good	Bad	Acceptable	Acceptable
Other parameters	*Large*	*Large*	*Large*	*Moderate*
Confluence effect	Bad	Bad	Very Bad	Acceptable
Detention effect	Very Bad	Very Bad	Very Bad	Acceptable

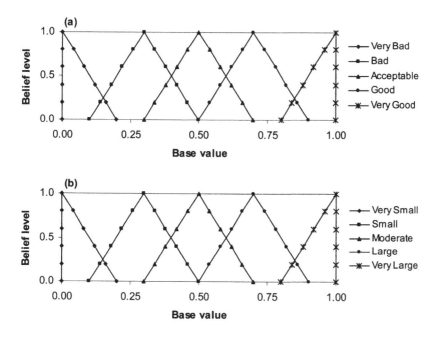

Figure 6.13. Membership functions of linguistic variables used for the qualitative evaluation of parameters with respect to (a) *Quality* and (b) *Importance*.

6.4.2 Analysis procedure and results

The computational detail of the expert judgement-based qualitative method is presented in Subsection 3.2.2. In summary the calculation steps are as follows:

1. Derive the *Quality* for each parameter from the *Qualities* of the sub-parameters, Equation (3.22).

2. Compute the uncertainty contribution of each of the parameters to the uncertainty in the output, Equation (3.24).

3. Estimate the total uncertainty in the forecast due to the uncertainty parameters, Equation (3.25).

4. Repeat Steps 1 to 3 for each expert.

5. Compute the average total uncertainty for all the experts, Equation (3.26).

6. Repeat Steps 1 to 5 to evaluate the *best-case* scenario with the Qualities of all the parameters as *Very Good*.

7. Repeat Steps 1 to 5 to evaluate the *worst-case* scenario with the Qualities of all the parameters as *Very Bad*.

8. Develop Fuzzy Qualitative and Crisp Qualitative Scales using the procedures as described in Section 4.3.

The uncertainty estimates in the output (represented by membership functions) computed from the evaluations made by each expert are presented in Figure 6.14. The average of the four experts is also shown in the figure. Table 6.4 summaries the differences (in terms of the defuzzified values) in the evaluations by the different experts. The defuzzification, which is a process of representing a fuzzy set by a single crisp value, is carried out using the centre-of-area method of defuzzification (Appendix I).

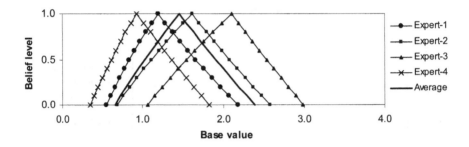

Figure 6.14. Output uncertainty computed form the individual assessments of experts on the quality and importance of the parameters.

Table 6.4: Comparison of uncertainty evaluations by different experts.

Evaluations	Defuzzified value	% difference from the average
Expert-1	1.31	13.2
Expert-2	1.62	7.9
Expert-3	2.05	36.3
Expert-4	1.04	31.1
Average	1.51	--

The results from Expert-1 and Expert-2 are very close to the average result, whereas, the results of Expert-3 and Expert-4 show significant deviations from the average.

The Fuzzy Qualitative and Crisp Qualitative Scales based on the *best-case* and the *worst-case* are shown in Figure 6.15(a) and (b), respectively. The scale consists of five divisions: *Very Small Uncertainty* (VSU), *Small Uncertainty* (SU), *Moderate Uncertainty* (MU), *Large Uncertainty* (LU) and *Very Large Uncertainty* (VLU) with the two extremes VSU and VLU corresponding respectively to the *best-case* and the *worst-case* scenario results. The evaluated average uncertainty and the uncertainty evaluated based on the assessments of Expert-3 and Expert-4 are presented on the scales.

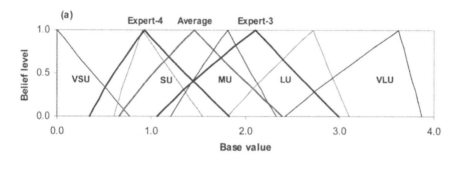

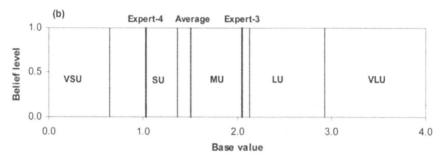

Figure 6.15. Estimated uncertainty based on the assessments by Expert-3, Expert-4 and the average of the four experts represented on: (a) fuzzy qualitative scales (b) crisp qualitative scale. The thin lines represent the regions of uncertainty scales and the thick lines represent the assessed uncertainty.

6.4.3 Conclusions

The qualitative method of uncertainty quantification presented here is relatively fast and easy. Uncertainties due to all identified sources can be incorporated in a relatively short time and with a small burden of computation. Moreover, experts' judgements over *Quality* and *Importance* can be updated with time by learning from experiences of the past, and thereby improving the quality of the uncertainty estimation. An obvious weakness of this qualitative method is that it can be very approximate and rather holistic. Another drawback of this approach observed in the present example is that the evaluations by different experts may differ noticeably. The results would become more credible if the experts' evaluation could be checked against either measurement or results from other methods. Besides providing such a check, the joint application of the quantitative and qualitative methods can also provide additional information regarding uncertainty.

6.5 Conclusions and discussion

This chapter presented the application of the FOSM, IFOSM and fuzzy set theory and expert judgement-based qualitative method for the assessment of uncertainty in the forecasts from Loire River flood forecasting model. Conclusions specific to each of the methods are presented at the end of each section describing the application of the methods. The conclusions presented here concern mainly the comparative advantages and limitations of these methods.

The popular FOSM method proved very useful and worked well in most flow conditions for the estimation of uncertainty in the forecasts from the Loire River model. It failed however in some specific flow conditions, where the IFOSM method has been used successfully. Selection of the appropriate size of the perturbation ratio is an important issue in the application of both the FOSM and IFOSM methods. In flood forecasting problems it is advisable to fix beforehand a nominal size of the perturbation ratio with the aid of a MC-type method for different flow conditions.

The main advantage of the expert judgement-based qualitative method is the possibility of including all sources of uncertainty without much computational burden. But the results are only qualitative and rather approximate. Assessments by experts are also difficult to obtain. In flood forecasting problems, the quantitative methods, such as MC, FOSM and fuzzy EP should be preferred where practicable. Results of the qualitative methods can however be used as an additional information.

Chapter 7

CONCLUSIONS AND RECOMMENDATIONS

Summary of Chapter 7

This research has been devoted to providing a contribution to the development of a framework and tools and techniques for uncertainty modelling in flood forecasting systems, particularly using probability and fuzzy set theories. It has also contributed to the exploration of hybrid techniques for uncertainty modelling and probability-possibility transformations. Principal conclusions drawn from this research and a set of recommendations for future research and development in the field of uncertainty modelling in flood forecasting and warning systems are presented in this chapter. The first conclusion addresses the issue of uncertainty in general in flood forecasting. The second conclusion is about the theories and methods for uncertainty modelling, which in particular highlights the analogies and differences between the popular Monte Carlo method and the fuzzy Extension Principle. The rest of the conclusions are specific to the methods and methodologies developed and/or applied and explored in this research. The recommendations emphasise on the need of uncertainty assessment in flood forecasting and advice that the probabilistic and possibilistic approaches should be viewed as complementary. The implementation of risk-based flood warnings is recommended and research needs for the hybrid techniques of uncertainty modelling and for the consideration of uncertainty in data-driven models are pointed out.

7.1 Conclusions

7.1.1 Uncertainty in flood forecasting systems

Flooding, like all natural disasters, is a complex and inherently uncertain phenomenon. In flood forecasting using mathematical models, the sources of uncertainty have been classified as (i) model uncertainty, (ii) input uncertainty, (iii) parameter uncertainty, and (iv) natural and operational uncertainty. In spite of the increasing advancement in the development of flood forecasting models and techniques, uncertainty in flood forecasts remains unavoidable. It is therefore important that the existence of uncertainty be admitted and properly appraised. Hiding uncertainty may create the illusion of certainty, the consequences of which can be very large (Krzysztowicz, 2001a). Various benefits of estimating uncertainty in flood forecasting have been identified, which include the rational basis for flood warning (risk-based warnings) and potential economic benefits from flood forecasting and warning systems. Even with the lack of risk-based flood warning procedures, quantifying uncertainty provides additional information about the forecasts and helps decision makers to use their own judgement more appropriately.

7.1.2 Theories and methods for modelling uncertainty

In uncertainty representation and modelling, probability theory and relatively recently fuzzy set theory (including fuzzy measures or possibility theory) have had the widest application. In flood forecasting, however, the application of theories other than probability are so far insignificant. The applications of fuzzy set theory in other fields of engineering have demonstrated its potential in modelling uncertainty (Schulz and Huwe, 1997 & 1999; Guyonnet et al., 1999; Revelli and Ridolfi, 2002). This research has extended the use of fuzzy set theory in flood forecasting. The successful implementation of the uncertainty assessment methodology using temporal disaggregation of time series inputs developed during this research showed the potential applicability of the fuzzy set approach in flood forecasting.

The probability theory-based methods like MC and FOSM are the most widely used and useful tools for uncertainty modelling. The use of the fuzzy Extension Principle (fuzzy set theory-based method) is also growing. The FOSM method propagates only the central values of the uncertain variables and requires less computational time. Unlike the FOSM method, both the MC method and the fuzzy EP propagate the complete functions representing the uncertainty in the variables. The computational requirements of the MC method and the EP are also comparable. The differences between the MC method and the EP are also significant. In the MC method, the scenarios that combine low probability parameter values have less chance of being randomly selected; whereas, in the EP all possible combinations of parameter values are considered, and the maximum and minimum model outputs obtained for the given intervals of the parameter values are directly reflected in the output uncertainty. Therefore, the EP is by and large more conservative than the MC method. This suggests that the former is desirable when the extreme values corresponding to the parameter values at the tails of their distributions are important. Another important distinction between the MC method and the EP is the correlation between the parameters. The MC method allows the effects of correlation between the parameters to be accounted for. In contrast, the current state of knowledge about the EP does not allow the incorporation of the effect of the correlation between the input parameters.

7.1.3 Disaggregation of time series inputs for uncertainty assessment

As a part of this research, a methodology is developed which uses temporal disaggregation of time series inputs for uncertainty assessment. This methodology explicitly considers the uncertainty in the time series inputs in the form of a probability distribution or fuzzy membership function. The important characteristics of this approach are that (i) it can be used with both MC method (for the probabilistic approach) and with the EP (for the fuzzy approach), and (ii) it is independent of the structure of the forecasting model. In other words, it can be used with any rainfall-runoff-routing type of deterministic model. The methodology is applied to the flood forecasting model of Klodzko catchment with the precipitation time series as uncertain inputs. The results also show that the output uncertainty due to the uncertainty in the temporal distribution can be significantly dominant over the uncertainty due to the uncertainty in the magnitude of the precipitation. This suggests that using time-averaged precipitation over the catchment may lead to erroneous forecasts. For the implementation of this methodology three important issues were identified: (i) the generation of the disaggregation coefficients, (ii) the selection of the number of subperiods, and (iii) the simplification of the methodology. This study also attempted to provide some answers to these issues.

7.1.4 Use of genetic algorithms with fuzzy Extension Principle

The application of the EP to a non-monotonic function requires an algorithm for the determination of the maximum and minimum of the function values. In the implementation of the uncertainty assessment methodology using temporal disaggregation in the framework of the EP, genetic algorithms are used for the determination of maxima and minima. The global optimisation algorithms like GAs are particularly useful when commercial off-the-shelf (black-box) software is used for building forecasting models (Maskey et al., 2002b). The application results with two versions of GAs (conventional or normal and micro) show a good potential of these algorithms to combine with the fuzzy EP for the propagation of uncertainty. It is however advisable, where affordable, to evaluate the performances of several algorithms and to find the most suitable algorithm for the given problem.

7.1.5 FOSM and Improved FOSM methods

The FOSM method is one of the widely used methods in uncertainty modelling. It is simple in application, has less computational requirement than other methods and is particularly useful when the information about detailed distributions of the uncertainty parameters is not available. Like any other method, it suffers however from some limitations (Melching and Yoon, 1996). As part of this research, an improvement to the FOSM method is applied using the second-degree reconstruction of the function to be modelled. The Improved FOSM method has a particular advantage when the average value of the input variable corresponds to maximum/minimum values or to regions where the slope of the function is very mild compared to the effects of curvature (non-linearity). It is also important to note that the improved method retains the simplicity and the smaller computational requirement of the FOSM method.

The sensitivity of both the FOSM and the IFOSM methods to the size of the perturbation were analysed. The analysis carried out with the flood forecasting model (Chapter 6) showed that the best value of the perturbation ratio (PR) is in the range from 0.75 to 1.5 for the FOSM method and from 1.25 to 1.75 for the IFOSM method. The results suggest that it is advisable to use some other standard method, for example the MC method, to fix beforehand the best value of the perturbation for a range of possible scenarios.

7.1.6 Expert judgement-based method and Qualitative Uncertainty Scales

The expert judgement-based qualitative method for uncertainty assessment (Subsection 3.2.2) is applied to the flood forecasting model of River Loire. Given the qualitative assessments of the parameter uncertainty by the experts (with respect to quality and importance of the parameters), this method allows for the estimation of the uncertainty due to all recognisable sources without a significant increase in computation time. Although this method has the same mathematical structure as that of the FOSM method (Subsection 3.4.2), the evaluations of the quality and importance (similar to the variance and sensitivity, respectively, of the FOSM method) are fully based on expert judgement. Consequently, it is a very approximate and rather holistic method. Therefore, in flood forecasting, this method should be used only as a complement to the quantitative methods like MC, FOSM and EP, particularly to obtain some indication of the relative effects of some of the sources of uncertainty, which cannot be incorporated into the framework of the quantitative methods.

Also of importance is the way in which the results of this method are interpreted. Sundararajan (1994 & 1998) attempted to interpret the results of this method by comparing it with the results from the probabilistic method using the so-called bench-marking values. The information contained by the results of such a qualitative method may be better communicated if they are interpreted qualitatively. Therefore, in the course of this research, the Qualitative Uncertainty Scales were developed (Section 4.3) with which the results can be measured qualitatively.

7.1.7 Probability-possibility transformation and hybrid technique for uncertainty modelling

A hybrid technique of uncertainty modelling is defined as a method that makes use of both probabilistic and possibilistic or fuzzy approaches together. There are at least two types of situations where the hybrid technique can be useful. These situations are referred to in this thesis as Type I and Type II Problems. The Type I Problem is the situation where an uncertain variable possesses components from both randomness and fuzziness. The Type II Problem refers to the situation where there exists some parameters with enough data to characterise their uncertainty using probability distributions and some other uncertain parameters with very little or no data to characterise their uncertainty in a probabilistic manner. This latter set of parameters can be more suitably characterised for uncertainty by means of the possibilistic approach using fuzzy membership functions. This then gives rise to a system with two sets of uncertain parameters: one represented by probability distributions and the other represented by membership functions.

The major obstacle in modelling uncertainty of a system of Type II Problem comes from the differences of operation between random-random (R-R) and fuzzy-fuzzy (F-F) variables. Part of this research is also devoted to explore the differences and similarities in the operations between R-R and F-F variables, and to investigate a probability-possibility (or fuzzy) transformation that takes these differences into account. This research shows that the addition and multiplication of two fuzzy variables by the EP using the α-cut method is similar to corresponding operations between two functionally dependent random variables for some specific conditions. The transformations also provide an alternative method for the evaluation of the Extension Principle for a monotonic function without using the α-cut method. It appears that the direct implication of this finding is very limited, but it is certainly an important basis for further research into probability-possibility transformations and hybrid techniques of uncertainty modelling.

7.2 Recommendations for further research and development

7.2.1 Uncertainty assessment should be an integral component of flood forecasting systems

Despite the fact that the presence of uncertainty in flood forecasting is significant, the implementation of uncertainty analysis in operational real time flood forecasting is very limited. As long as forecasts are subject to uncertainty the only honest and sensible way is to admit it and to incorporate this information in decision making as much as possible. Hiding it creates the illusion of certainty, the consequences of which could be very large. Therefore, it is essential that all forecasting systems should have an uncertainty assessment procedure as an integral component.

7.2.2 Probabilistic and possibilistic approaches should be considered as complementary

The founder of fuzzy set theory, L.A. Zadeh himself, suggested that probabilistic and possibilistic approaches should be considered as complementary and NOT competitive (Zadeh, 1995). A similar view was reported by several other researchers, e.g. Kikuchi and Pursula (1998). The present author also believes that these two theories should be viewed as complementary. If this view is realised by the researchers and practitioners of uncertainty modelling, the art of uncertainty representation and modelling will certainly benefit.

7.2.3 Towards hybrid techniques of uncertainty modelling

The hybrid technique of uncertainty modelling, whereby the combined application of the probabilistic and possibilistic or fuzzy approaches is sought, has been discussed by several researchers. The fact that both the probabilistic and fuzzy set approaches have some limitations in representing and modelling uncertainty, it can be argued that the analysis of the uncertainty can be best handled using a combination of probabilistic and fuzzy approaches. Such a hybrid technique may have an advantage in incorporating both randomness and vagueness types of uncertainty explicitly. Therefore, to obtain the benefit from both the probability and fuzzy techniques, the research on the hybrid technique of modelling uncertainty should be explored further.

In some situations of uncertainty/risk-based decision making, for example for flood warning, the application of the hybrid technique may be appropriate. The risk-based decision-making requires that the risks be evaluated for different decision alternatives. In probabilistic sense, *risk* is defined as the probability of occurrence of an (future) event times the consequence of the event. In the context of flood warnings, the future event is the anticipated flood and the consequence is the damages (also known as disutilities) of the flood. The uncertainty in the forecasted flood is commonly expressed probabilistically. The estimation of damages due to flooding is a rather complicated issue and generally involves the lack of data for the probabilistic quantification of uncertainty. In this case the hybrid concept, called *fuzzy-probabilistic risk* may be used whereby the uncertainty in the estimated damage is characterised by a fuzzy membership function based on limited available information Maskey (2004). The author anticipates growing applications of such an approach and believes that enough attention should be given to further explore such concepts in future research.

7.2.4 Towards uncertainty and risk-based flood forecasting and warning systems

Risk-based design of civil engineering structures is becoming increasingly common (see Tung and Mays, 1981; Van Gelder, 2000; Voortman, 2002; Vrijling 1993; Vrijling et al., 1998). Well formulated mathematical methods are available for risk based flood warnings (Krzysztofowicz, 1993). The risk-based warnings are more rational, offer economic benefits and their needs are being increasingly realised (Kelly and Krzysztofowicz, 1994; Krzysztofowicz, 1993; Krzysztofowicz et al., 1994). But their widespread implementation in practice is yet to be witnessed. Therefore the implementation of such a system should be encouraged whenever practicable to achieve additional economic benefits from flood forecasting and warning systems.

7.2.5 Uncertainty assessment in data-driven modelling

Some of the data-driven techniques, such as the fuzzy rule-based systems and the fuzzy regression, work with imprecise data and implicitly incorporate the uncertainty concept in modelling. These models however do not have the flexibility of using uncertainty analysis methods based on other theories (e.g. the most popular probability theory). In the case of an ANN-based model (most popular so far of the data-driven techniques), model performance is, by and large, based on the difference between the observed and model predicted results using the measures like Root Mean Square Error (RMSE). Methods for the application of more commonly used uncertainty techniques, such as based on probability and fuzzy set theories should be explored for the ANN-based and other data-driven models so that these models can also be used in decision making based on uncertainty and risk.

Appendix I

FUZZY SETS, FUZZY ARITHMETIC AND DEFUZZIFICATION

I.1 Definitions on fuzzy sets

Fuzzy set theory has been introduced in Subsection 2.3.2. Definitions on some more terminologies used in the fuzzy set theory are presented here. More detailed coverage of this topic can be found in Bardossy and Duckstein (1995), Dubois and Prade (1988), Kaufmann and Gupta (1991), Ross (1995) and Zimmermann (1991).

Fuzzy set

Let X be a universe set of x values (elements). Then A is called a fuzzy (sub)set of X, if A is a set of ordered pairs:

$$A = \{(x, \mu_A(x)), x \in X, \mu_A(x) \in [0,1]\} \tag{I.1}$$

where $\mu_A(x)$ is the grade of membership (or degree of belief) of x in A. The function $\mu_A(x)$ is called the *membership function* of A.

Height of a fuzzy set

The *height* of a fuzzy set is the maximum value of the membership in its membership function, i.e. the height of the fuzzy set A

$$h(A) = \max\{\mu_A(x_1), ..., \mu_A(x_n), x \in X\} \tag{I.2}$$

Support of a fuzzy set

The *support* of the membership function of a fuzzy set is the region of the universe that is characterised by nonzero membership in the fuzzy set, i.e.

$$Supp(A) = \{x \mid \mu_A(x) > 0\} \tag{I.3}$$

Normal fuzzy set

A fuzzy set which has at least one element with unity membership is called a *normal* fuzzy set, that is, there exists $\mu_A(x) = 1$ for some $x \in X$. In terms of the height of the fuzzy set, a normal fuzzy set has $h(A) = 1$. If the height of a fuzzy set is less than unity the fuzzy set is called a subnormal.

Convex fuzzy set

A *convex* fuzzy set is the set in which the membership function is monotonically increasing or decreasing. Precisely, the membership function in a convex fuzzy set is either (i) monotonically increasing, or (ii) monotonically decreasing, or (iii) monotonically increasing and monotonically decreasing with increasing values of the elements in the universe. For every real number, say x_1, x_2 and x_3, with $x_1 < x_2 < x_3$

$$\mu_A(x_2) \geq \min\{\mu_A(x_1), \mu_A(x_3)\} \quad\quad\quad\quad (\text{I.4})$$

Fuzzy number

A fuzzy set which is normal as well as convex is called a *fuzzy number*.

Fuzzy Extension Principle

Consider a function of several uncertain variables:

$$y = f(x_1, \dots, x_n) \quad\quad\quad\quad (\text{I.5})$$

Let fuzzy sets $A_1, \dots, A_n$ be defined on the universes $X_1, \dots, X_n$ such that $x_1 \in X_1, \dots, x_n \in X_n$. The mapping of these input sets can be defined as a fuzzy set B:

$$B = f(A_1, \dots, A_n) \quad\quad\quad\quad (\text{I.6})$$

where the membership function of the image B is given by

$$\mu_B(y) = \begin{cases} \max\{\min[\mu_{A1}(x_1), \dots, \mu_{An}(x_n)]; y = f(x_1, \dots, x_n)\} \\ \quad \text{if } \exists(x_1, \dots, x_n) \in X_1 \times \dots \times X_n \text{ such that } f(x_1, \dots, x_n) = y \\ 0 \quad \text{otherwise} \end{cases} \quad (\text{I.7})$$

The Equation (I.7) is the mathematical expression for the Extension Principle (EP) of fuzzy sets. The above equation is defined for a discrete-valued function f. If the function f is continuous-valued then the *max* operator is replaced by the *sup* (supremum) operator (the supremum is the least upper bound).

Alpha-cut of a fuzzy set

An α-*cut* (alpha-cut) of a fuzzy set A, denoted as A^α is the set of elements x of a universe of discourse X for which the membership function of A is greater than or equal to α. That is,

$$A^{(\alpha)} = \left\{ x \in X, \mu_A(x) \geq \alpha, \alpha \in [0,1] \right\} \tag{I.8}$$

α-cut provides a convenient way of performing arithmetic operations on fuzzy sets and fuzzy numbers including in applying the EP. Let us consider a fuzzy number (a membership function) as shown in Figure I.1. Let a α-cut level intersects at two points a and b on the membership function. The values of the variable x corresponding to points a and b are x_1 and x_2 ($x_1, x_2 \in X$), respectively. Then the set A^α contains all possible values of the variable X including and between x_1 and x_2. The x_1 and x_2 are commonly referred as lower and upper bounds of the α-cut.

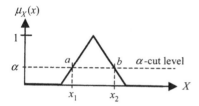

Figure I.1. Illustration of an α-cut of a fuzzy set

I.2 Fuzzy arithmetic

Fuzzy arithmetic is based on the EP introduced by Zadah (1975) and elaborated by Yager (1986). Application of the EP to a function of fuzzy variables, also called fuzzy numbers (Kaufmann and Gupta, 1991), has been explained in Subsection 3.2.1. The implementation of the EP can be distinguished for monotonic and non-monotonic functions. For a non-monotonic function its implementation requires an algorithm for the determination of maximum and minimum values of the function to be evaluated. For a monotonic function the EP can be greatly simplified as mentioned in Subsection 3.2.1. The EP for a monotonic function can be applied to perform the basic operations (addition, subtraction, multiplication and division) between two fuzzy numbers, which are commonly referred to as fuzzy arithmetic. This section presents the fuzzy arithmetic for some basic operations. Readers are referred to Kaufmann and Gupta (1991) for a comprehensive coverage of fuzzy arithmetic.

Let $\widetilde{X}$ and $\widetilde{Y}$ are two fuzzy numbers represented by membership functions as shown in Figure I.2. The fuzzy arithmetic consists in finding the upper bound (UB) and the lower bound (LB) of the resulting fuzzy number $\widetilde{Z} = \widetilde{X}(\circ)\widetilde{Y}$, where the operator $(\circ) \in [(+),(-),(\times),(:)]$ at any value of the α-cut ($\alpha \in [0,1]$). The lower and upper bound values of $\widetilde{X}$, $\widetilde{Y}$ and $\widetilde{Z}$ are shown in Figure I.2.

Addition

Let $\tilde{Z} = \tilde{X}(+)\tilde{Y}$. Then

$$z_{LB}^{(\alpha)} = \min\{(x_{LB}^{(\alpha)} + y_{LB}^{(\alpha)}),(x_{LB}^{(\alpha)} + y_{UB}^{(\alpha)}),(x_{UB}^{(\alpha)} + y_{LB}^{(\alpha)}),(x_{UB}^{(\alpha)} + y_{UB}^{(\alpha)})\}$$
$$= (x_{LB}^{(\alpha)} + y_{LB}^{(\alpha)}) \tag{I.9}$$

$$z_{UB}^{(\alpha)} = \max\{(x_{LB}^{(\alpha)} + y_{LB}^{(\alpha)}),(x_{LB}^{(\alpha)} + y_{UB}^{(\alpha)}),(x_{UB}^{(\alpha)} + y_{LB}^{(\alpha)}),(x_{UB}^{(\alpha)} + y_{UB}^{(\alpha)})\}$$
$$= (x_{UB}^{(\alpha)} + y_{UB}^{(\alpha)}) \tag{I.10}$$

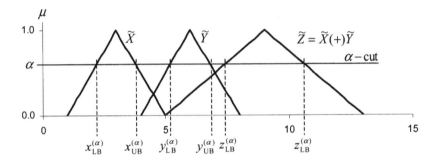

Figure I.2. Addition of two fuzzy numbers defined for any value of $\alpha = [0, 1]$.

Subtraction

Let $\tilde{Z} = \tilde{X}(-)\tilde{Y}$. Then

$$z_{LB}^{(\alpha)} = \min\{(x_{LB}^{(\alpha)} - y_{LB}^{(\alpha)}),(x_{LB}^{(\alpha)} - y_{UB}^{(\alpha)}),(x_{UB}^{(\alpha)} - y_{LB}^{(\alpha)}),(x_{UB}^{(\alpha)} - y_{UB}^{(\alpha)})\}$$
$$= (x_{LB}^{(\alpha)} - y_{UB}^{(\alpha)}) \tag{I.11}$$

$$z_{UB}^{(\alpha)} = \max\{(x_{LB}^{(\alpha)} - y_{LB}^{(\alpha)}),(x_{LB}^{(\alpha)} - y_{UB}^{(\alpha)}),(x_{UB}^{(\alpha)} - y_{LB}^{(\alpha)}),(x_{UB}^{(\alpha)} - y_{UB}^{(\alpha)})\}$$
$$= (x_{UB}^{(\alpha)} - y_{LB}^{(\alpha)}) \tag{I.12}$$

Multiplication

Let $\tilde{Z} = \tilde{X}(\times)\tilde{Y}$. Then

$$z_{LB}^{(\alpha)} = \min\{(x_{LB}^{(\alpha)} y_{LB}^{(\alpha)}),(x_{LB}^{(\alpha)} y_{UB}^{(\alpha)}),(x_{UB}^{(\alpha)} y_{LB}^{(\alpha)}),(x_{UB}^{(\alpha)} y_{UB}^{(\alpha)})\}$$
$$= (x_{LB}^{(\alpha)} y_{LB}^{(\alpha)}) \tag{I.13}$$

$$z_{UB}^{(\alpha)} = \max\{(x_{LB}^{(\alpha)} y_{LB}^{(\alpha)}),(x_{LB}^{(\alpha)} y_{UB}^{(\alpha)}),(x_{UB}^{(\alpha)} y_{LB}^{(\alpha)}),(x_{UB}^{(\alpha)} y_{UB}^{(\alpha)})\}$$
$$= (x_{UB}^{(\alpha)} y_{UB}^{(\alpha)}) \tag{I.14}$$

Division

Let $\widetilde{Z} = \widetilde{X}(:)\widetilde{Y}$. Then

$$
\begin{aligned}
z_{LB}^{(\alpha)} &= \min\{(x_{LB}^{(\alpha)}/y_{LB}^{(\alpha)}),(x_{LB}^{(\alpha)}/y_{UB}^{(\alpha)}),(x_{UB}^{(\alpha)}/y_{LB}^{(\alpha)}),(x_{UB}^{(\alpha)}/y_{UB}^{(\alpha)})\} \\
&= (x_{LB}^{(\alpha)}/y_{UB}^{(\alpha)})
\end{aligned}
\tag{I.15}
$$

$$
\begin{aligned}
z_{UB}^{(\alpha)} &= \max\{(x_{LB}^{(\alpha)}/y_{LB}^{(\alpha)}),(x_{LB}^{(\alpha)}/y_{UB}^{(\alpha)}),(x_{UB}^{(\alpha)}/y_{LB}^{(\alpha)}),(x_{UB}^{(\alpha)}/y_{UB}^{(\alpha)})\} \\
&= (x_{UB}^{(\alpha)}/y_{LB}^{(\alpha)})
\end{aligned}
\tag{I.16}
$$

Multiplication of a fuzzy number by a crisp number

The multiplication between fuzzy–fuzzy numbers (Eqs. (I.13) and (I.14)) can be extended to the multiplication between a crisp number and a fuzzy number. Let in $\widetilde{Z} = \widetilde{X}(\times)\widetilde{Y}$, $\widetilde{X}$ is a crisp number whose value is x^*, that is $\widetilde{X} = X = x^*$. Which means that

$$
\left.\begin{aligned}
\mu_X(x = x^*) &= 1 \\
\mu_X(x \neq x^*) &= 0
\end{aligned}\right\}
\tag{I.17}
$$

This then follows in Equations (I.13) and (I.14) that

$$
x_{LB}^{(\alpha)} = x_{UB}^{(\alpha)} = x^*
\tag{I.18}
$$

Substituting Equation (I.18) into Equations (I.13) and (I.14) we obtain

$$
\begin{aligned}
z_{LB}^{(\alpha)} &= \min\{(x^* y_{LB}^{(\alpha)}),(x^* y_{UB}^{(\alpha)}),(x^* y_{LB}^{(\alpha)}),(x^* y_{UB}^{(\alpha)})\} \\
&= x^* y_{LB}^{(\alpha)}
\end{aligned}
\tag{I.19}
$$

$$
\begin{aligned}
z_{UB}^{(\alpha)} &= \max\{(x^* y_{LB}^{(\alpha)}),(x^* y_{UB}^{(\alpha)}),(x^* y_{LB}^{(\alpha)}),(x^* y_{UB}^{(\alpha)})\} \\
&= x^* y_{UB}^{(\alpha)}
\end{aligned}
\tag{I.20}
$$

I.3 Defuzzification methods

There are various methods of *defuzzification*, whereby a fuzzy set is represented by a single crisp value. Two commonly used methods are the *centre-of-area method* and the *max-membership method*. In the former (also called *centroid method*) the defuzzified value x^* is defined as the value within the range of the base value for which the area under the graph of the membership function, $\mu_X(x)$, is divided into two equal sub-areas. This value is calculated by the formula

$$x^* = \frac{\int \mu_X(x)x\,dx}{\int \mu_X(x)\,dx} \qquad (I.21)$$

The latter method, also know as *height method*, seeks for the value corresponding to the maximum membership, i.e.

$$\mu_X(x^*) \geq \mu_X(x); \qquad x \in X \qquad (I.22)$$

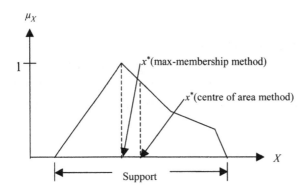

Figure I.3. A membership function showing the defuzzification values by *centre-of-area* method and the *max-membership* method.

An example of defuzzification by the method of *centre-of-area* and the method of *max-membership* are shown in Figure I.3. If the membership function is defined by a triangular shape, the defuzzification by the centre of area method reduces to a simple form as

$$x^* = \frac{x_1 + x_2 + x_3}{3} \qquad (I.23)$$

where the x_1, x_2 and x_3 are as shown in Fig. I.4.

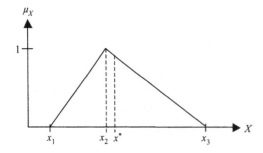

Figure I.4. Defuzzification of a triangular membership function by the *centre-of-area* method.

Abbreviations

ANN	: Artificial Neural Network
BFS	: Bayesian Forecasting System
CDF	: Cumulative Distribution Function
CN	: Curve Number
CoV	: Coefficient of Variation
EP	: Extension Principle
F-F	: Fuzzy-Fuzzy
FFWRS	: Flood Forecasting, Warning and Response System
FOSM	: First-Order Second Moment
GA	: Genetic Algorithm
GLUE	: Generalised Likelihood Uncertainty Estimation
GO	: Global Optimization
GOA	: Global Optimization Algorithm
HEC	: Hydraulic Engineering Centre
HMS	: Hydrologic Modelling System
IFOSM	: Improved First-Order Second Moment
LB	: Lower Bound
MC	: Monte Carlo
OSIRIS	: Operational Solutions for the Management of Inundation Risks in the Information Society
PDF	: Probability Density Function
PR	: Perturbation Ration
RMSE	: Root Mean Square Error
R-R	: Random-Random
SCS	: Soil Conservation Service
SHE	: European Hydrological System – Systeme Hydrologique European
UB	: Upper Bound

Notations

Notations that are used locally are excluded in this list.

$\forall$	: for all
$\exists$	: there exists
(+)	: addition between fuzzy numbers
(−)	: subtraction between fuzzy numbers
(×)	: multiplication between fuzzy numbers
(:)	: division between fuzzy numbers
α	: alfa-cut (α-cut) level in fuzzy membership (used as superscript)
γ	: coefficient of variation
σ	: standard deviation
ε	: perturbation
Φ	: quality of a parameter (in the form of a membership function), which is analogous to the variance of the parameter
Ψ	: importance of a parameter (in the form of a membership function), which is analogous to the sensitivity of the parameter
μ_X	: membership function of X
$\widetilde{A}, \widetilde{B}$	: fuzzy sets
$b_{i,j}$	: disaggregation coefficient for subbasin i, subperiod j
$Cov(X_i, X_j)$	: covariance between X_i, X_j
$H(p)$	: *Shannon entropy* (an uncertainty measure in probability theory)
LB	: lower bound (used as a subscript)
$N(r)$	: *nonspecificity* (an uncertainty measure in possibility theory)
P_X	: cumulative (probability) distribution function of X
p_X	: probability density function of X
r_X	: possibility distribution function of X
P_i	: accumulated precipitation depth for subbasin i.
$p_{i,j}$	: disaggregated precipitation depth for subbasin i, subperiod j.
$P(\widetilde{F})$	: probability of a fuzzy event $\widetilde{F}$
Q	: discharge
R	: hydraulic radius

$S(r)$	: *strife* (an uncertainty measure in possibility theory)
UB	: upper bound (used as subscript)
U_o	: total uncertainty in an output (in the form of a membership function)
$U_T(r)$	: total uncertainty measure in possibility theory
$Var(X)$	: variance of X
W_i	: value of any time series input for a given time period for subbasin i
$w_{i,j}$	: disaggregated value of any time series input for subbasin i, subperiod j
X	: uncertain variable
	: also a universal set
x	: a value of the uncertain variable X
$\bar{x}$	: mean value of x
$\tilde{X}$	: fuzzy number
$\tilde{Y}$	: fuzzy number
$\tilde{Z}$	: fuzzy number

References

Abbott, M.B., Bathurst, J.C., Cunge, J.A., O'Connell, P.E. and Rasmussen, J. (1986a). An introduction to the European Hydrological System – Systeme Hydrologique Europeen, "SHE", 1: History and philosophy of a physically-based, distributed modelling system. *Journal of Hydrology*, **87**, pp. 45-59.

Abbott, M.B., Bathurst, J.C., Cunge, J.A., O'Connell, P.E. and Rasmussen, J. (1986b). An introduction to the European Hydrological System – Systeme Hydrologique Europeen, "SHE", 2: Structure of a physically-based, distributed modelling system. *Journal of Hydrology*, **87**, pp. 61-77.

Ang, A. H-S. and Tang, W. H. (1975). Probability Concepts in Engineering Planning and Design, Volume I-Basic Principles. John Wiley & Sons, Inc.

Aronica, G., Hankin, B. and Beven, K. (1998). Uncertainty and Equifinality in Calibrating Distributed Roughness Coefficients in a Flood Propagation Model with Limited Data. *Advances in Water Resources*, **22**(4), pp. 349-365.

Ayyub, B.M. (2001). Elicitation of Expert Opinions for Uncertainty and Risks. CRC Press.

Ayyub, B. and Chao, R.-J. (1998). Uncertainty Modelling in Civil Engineering with Structural and Reliability Applications. In *Uncertainty Modelling and Analysis in Civil Engineering*, Ayyub, B.M. [ed.], CRS Press, pp. 3-32.

Babovic, V. and Keijzer, M. (2000). Genetic programming as a model induction engine. *Journal of Hydroinformatics*, **2**(1), pp. 35-60.

Bardossy, A., Bogardi, I. and Duckstein, L. (1990). Fuzzy Regression in Hydrology. *Water Resources Research*, **26**(7), pp. 1497-1508.

Bardossy, A. and Duckstein, L. (1995). Fuzzy Rule-Based Modelling with Applications to Geophysical, Biological and Engineering Systems. CRC Press.

Beven, K.J. (1989). Changing ideas in hydrology – the case of physically-based models. *Journal of Hydrology*, **105**, pp. 157-172.

Beven, K.J. and Binley, A. (1992). The future of distributed models: model calibration and uncertainty prediction. *Hydrological Processes*, **6**, pp. 279-298.

Beven, K. and Freer, J. (2001). Equifinality, data assimilation, and uncertainty estimation in mechanistic modelling of complex environmental systems using the GLUE methodology. *Journal of Hydrology*, **249**, pp. 11-29.

Beven, K., Lamb, R., Quinn, P., Romanowicz, R., and Freer, J. (1995). TOPMODEL. In *Computer Models of Watershed Hydrology*, Singh, V.P. (ed.), Water Resources Publication, Colorado, pp. 627-668.

Beven, K.J. and Kirkby, M.J. (1979). A physically-based variable contributing area model of basin hydrology. *Hydrological Sciences Bulletin*, **24**(1), pp. 43-69.

Blancher, P., Cabal, A., Delahaye, A. and Xhaard, H. (2003). Societal Expectations from and Preparedness to ICT Based Flood Risk Management. In *Proc., OSIRIS Workshop: Flood Events: Are We Prepared*, March 2003, Berlin, Germany, pp. 51-64.

Borga, M. (2002). Accuracy of radar rainfall estimates for streamflow simulation. *Journal of Hydrology*, **267**, pp. 26-39.

Box, G.E.P. and Tiao, G.C. (1973). *Bayesian Inference in Statistical Analysis*. Addison-Wesley, USA.

Burian, S.J. and Durrans (2002). Evaluation of an artificial neural network rainfall disaggregation model. *Water Science and Technology*, **45**(2), pp. 99-104, IWA Publishing.

Burnash, R.J.C., Ferrel, R.L. and McGuire, R.A. (1973). *A general streamflow simulation system – conceptual modelling for digital computers*. Report by the Joint Federal State River Forecasting Centre, Sacramento, CA.

Carpenter, T.M., Georgakakos, K.P. and Sperfslagea, J.A. (2001). On the parametric and NEXRAD-radar sensitivities of a distributed hydrologic model suitable for operational use. *Journal of Hyrdology*, **253**, pp. 169-193.

Carroll, D.L. (1996). Chemical laser modelling with genetic algorithms. *American Institute of Aeronautics and Astronautics (AIAA) Journal*, **34**(2), pp. 338-346.

Caselton, W.F. and Luo, W. (1992). Decision Making with Imprecise Probabilities: Dempster-Shafer Theory and Application. *Water Resources Research*, **28**(12), pp. 3071-3083.

Chang, C.-H., Tung, Y.-K. and Yang, J.-C. (1994). Monte Carlo Simulation for Correlated Variables with Marginal Distributions. *Journal of Hydraulic Engineering*, ASCE, **120**(3), pp. 313-331.

Chen, H.-K., Hsu, W.-K. and Chiang, W.-L. (1998). A comparison of vertex method with JHE method. *Fuzzy Sets and Systems*, **95**, pp. 201-214.

Chow, V.T., Maidment, D.R., and Mays, L.W. (1988). *Applied Hydrology*. McGraw-Hill, New York, NY.

Cluckie, I.D. and Collier, C.G. (1991). *Hydrological Applications of Weather Radar*. Ellis Horwood Ltd., England.

Collotte, P., Erlich, M. and Weets, G. (2003). OSIRIS Project Objectives and Achievements. In *Proc., OSIRIS Workshop: Flood Events: Are We Prepared*, March 2003, Berlin, Germany, pp. 7-15.

Crawford, N.H., and Linsley, R.K. (1966). *Digital simulation in hydrology: Standford Watershed Model Mark IV*. Dept. of Civil Engineering, Tech. Rep. **39**, Standford University, Standford.

Cullen, A.C. and Frey, H.C. (1999). Probabilistic techniques in exposure assessment: A hand book for dealing with variability and uncertainty in models and inputs. Plenum Press, New York.

Cunge, J.A., Holly, F.M. and Verwey, A. (1980). *Practical Aspects of Computational River Hydraulics*. Pitman Publishing Limited.

Dartmouth Flood Observatory (2003). Global Active Archive of Large Flood Events. Retrieved 10 October 2003 http://www.dartmouth.edu/%7Efloods/Archives/index.html.

Dawson, C.W. and Wilby, R. (1998). An artificial neural network approach to rainfall-runoff modelling. *Hydrological Sciences Journal*, **43** (1), pp. 47-66.

De Finetti, B. (1974). Theory of Probability: A critical introductory treatment, Volume 1. John Wiley & Sons.

Demspster, A.P. (1969). Upper and lower probability inferences for families of hypotheses with monotone density rations. *Annals of Mathematical Statistics*, 40(3), pp. 953-969.

Dibike, Y.B. (2001). Model induction from data: Towards the next generation of computational engines in hydraulics and hydrology. Swets & Zeitlinger B.V., Lisse.

Dubois, D. and Prade, H. (1980). *Fuzzy sets and systems: theory and application*. San Diego, Academic.

Dubois, D. and Prade, H. (1988). Possibility Theory: An Approach to Computerized Processing of Uncertainty. Plenum Press, New York, USA.

Dubois, D. and Prade, H. (1991). Random sets and fuzzy interval analysis. *Fuzzy Sets and Systems*, **42**, pp. 48-65.

Fattorelli, S., Dalla Fontana, G. and Da Ros, D. (1999). Flood Hazard Assessment and Mitigation. In *Floods and Landslides: Integrated Risk Assessment*, Casale, R. and Morgottini, C. (eds.), Springer, pp. 19-38.

Ferson, S. and Kuhn, R. (1994). Interactive microcomputer software for fuzzy arithmetic. In *Proceedings of High Consequence Operations Safety Symposium*, Sandia National Laboratories, SAND94-2364, Cooper, J.A. (ed.), Albuquerque, New Mexico, pp. 493-506.

Ferson, S. and Ginzburg, L. (1995). Hybrid Arithmetic. In *Proceedings of ISUMA-NAFIPS'95*, Bimal M. Ayyub (ed.), IEEE Computer Society Press, Los Alamitos, California, pp. 619-623.

Fishwick, P.A. (1991). Fuzzy simulation: specifying and identifying qualitative methods. *International Journal of General Systems*, **19**, pp. 295-316.

Fleming, G. (1975). *Computer simulation techniques in hydrology*. Elsevier Science, New York.

Gates, T.K., Alshaikh, A.A., Ahmed, S. I. and Molden, D.J. (1992). Optimal Irrigation Delivery System Design Under Uncertainty. *Journal of Irrigation and Drainage Engineering*, ASCE, **118**(3), pp. 433-449.

Gautam, D.K. and Holz, K.P. (2001). Rainfall-runoff modelling using adaptive neuro-fuzzy systms. *Journal of Hydroinformatics*, **3**(1), pp. 3-10.

Gen, M. and Cheng, R. (2000). *Genetic algorithms and engineering optimization*. A Wiley-Interscience Publication, John Wiley & Sons, Inc.

Glassheim, E. (1997). Fear and loathing in North Dakota. *Natural Hazards Observer*, **XXI** (6), pp. 1-4.

Goldberg, D.E. (1989). *Genetic algorithms in search, optimization, and machine learning*. Addison-Wesley Publishing Company, Inc.

Goldberg, D.E. and Richardson, J. (1987). Genetic algorithms with sharing for multimodal function optimization. In *Genetic algorithms and their applications: Proceedings of the Second International Conference on Genetic Algorithms*, pp. 41-49.

Goodwin, P. and Hardy, T.B. (1999). Integrated simulation of physical, chemical and ecological processes for river management. *Journal of Hydroinformatics*, **1**(1), pp. 33-58.

Grijsen, J.G., Snoeker, X.C., Vermeulen, C.J.M., El Amin, M. and Mohamed, Y.A. (1992). An Information System for Flood Early Warning. In *Floods and Flood Management*, Saul, A.J. (ed.), pp. 263-289.

Grimmett. G.R. and Stirzaker, D.R. (1988). *Probability and Random Processes*. Oxford University Press, Oxford, UK.

Guinot, V. (1998). Sensitivity of Model Outputs to Parameters. In *Proceedings of Hydroinformatics 98*, Copenhagen, Vladan, B. and Larsen, L.C. (eds.), A.A. Balkema, Rotterdam, pp. 1095-1100.

Guinot, V. and Gourbesville, P. (2003). Calibration of physically based models: back to basics. *Journal of Hydroinformatics*, **5**(4), pp. 233-244.

Guyonnet, D., Come, B., Perrochet, P. and Parriaux, A. (1999). Comparing two methods for addressing uncertainty in risk assessments. *Journal of Environmental Engineering*, **125**(7), pp. 660-666.

Haldar, A. and Mahadevan, S. (2000a). Probability, Reliability, and Statistical Methods in Engineering Design. John Wiley & Sons, Inc.

Haldar, A. and Mahadevan, S. (2000b). *Reliability Assessment Using Stochastic Finite Element Analysis*. John Wiley & Sons, Inc.

Hall, J. W. (1999). *Uncertainty Management for Coastal Defence Systems*. PhD Thesis, University of Bristol, Department of Civil Engineering, UK.

Holland, J.H. (1975). *Adaptation in natural and artificial systems*. The University of Michigan Press.

Johnson, P. A. and Ayyub, B. M. (1996). Modelling Uncertainty in Prediction of Pier Scour. *Journal of Hydraulic Engineering*, ASCE, **122**(2), pp. 66-72.

Kaufmann, A. and Gupta, M.M. (1991). *Introduction to Fuzzy Arithmetic, Theory and Applications*. Van Nostrand Reinhold.

Kelly, K.S. and Krzysztofowicz, R. (1994). Probability distribution for flood warning systems. *Water Resources Research*, **30**(4), pp. 1145-1152.

Kelly, K.S. and Krzysztofowicz, R. (2000). Precipitation Uncertainty Processor for Probabilistic River Stage Forecasting. *Water Resources Research*, **36**(9), pp. 2643-2653.

Kikuchi, S. and Pursula, M. (1998). Treatment of Uncertainty in Study of Transportation: Fuzzy Set Theory and Evidence Theory. *Journal of Transportation Engineering*, ASCE, **124**(1), pp. 1-8.

Kitanidis, P.K. and Bras, R.L. (1980). Real-time forecasting with a conceptual model, 1. Analysis of Uncertainty. *Water Resources Research*, **16**(6), pp. 1025-1033.

Klir, G.J. (1992). Probabilistic versus possibilistic conceptualisation of uncertainty. *Analysis and Management of Uncertianty: Theory and Application*, Ayyub, B.M., Gupta, M.M. and Kanal, L.N. (eds.), Elsevier Science Publishers, pp. 13-25.

Klir, G.J. and Folger, T.A. (1988). *Fuzzy Sets, Uncertainty and Information*. Prentice Hall International, London.

Klir G.J. and Wierman, M.J. (1998). *Uncertainty-Based Information*, Physica-Verlag New York.

Klir, G.J. and Yuan, B. (1995). *Fuzzy Sets and Fuzzy Logic, Theory and Applications.* Prentice Hall PTR.

Koutsoyiannis, D. and Onof, C. (2001). Rainfall diasggregation using adjusting procedures on a Poisson cluster model. *Journal of Hydrology*, **246**, pp. 109-122.

Krajewski, W.F. and Smith, J.A. (1995). Uncertainty analysis in radar-rainfall estimation. In *New Uncertainty Concepts in Hydrology and Water Resources*, Kundzewicz, Z.A. (ed.), Cambridge University Press, Cambridge, UK, pp. 181-189.

Krisnakumar, K. (1989). Micro-genetic algorithms for stationary and non-stationary function optimization. *SPIE: Intelligent Control and Adaptive Systems*, 1196, Philadelphia, PA.

Krzysztofowicz, R. (1993). A Theory of Flood Warning Systems. *Water Resources Research*, **29**(12), pp. 3981-3994.

Krzysztofowicz, R. (1995). Recent advances associated with flood forecast and warning systems. *Reviews of Geophysics*, Vol. **33** Supplement, American Geophysical Union. Retrieved 14 March 2000 http://earth.agu.org/revgeophys/krzysz00/krzysz00.html.

Krzysztofowicz, R. (1999). Bayesian theory of probabilistic forecasting via deterministic hydrologic model. *Water Resources Research*, **35**(9), pp. 2739-2750.

Krzysztofowicz, R. (2001a). The case for probabilistic forecasting in hydrology. *Journal of Hydrology*, **249**, pp. 2-9.

Krzysztofowicz, R. (2001b). Integrator of uncertainties for probabilistic river stage forecasting: precipitation-dependent model. *Journal of Hydrology*, **249**, pp. 69-85.

Krzysztofowicz, R. and Davis, D.R. (1983). A methodology for evaluation of flood forecast-response systems, 1, Analysis and concepts. *Water Resources Research*, **19**(6), pp. 1423-1429.

Krzysztofowicz, R. and Kelly, K.S. (2000). Hydrologic uncertainty processor for probabilistic river stage forecasting. *Water Resources Research*, **36**(11), pp. 3265-3277.

Krzysztofowicz, R., Kelly, K.S. and Long, D. (1994). Reliability of Flood Warning Systems. *J of Water Resources Planning and Management*, ASCE, **120**(6), 906-926.

Kumar, D.N., Lall, U. and Petersen, M.R. (2000). Multisite disaggregation of monthly to daily streamflow. *Water Resources Research*, **36**(7), pp.1823-1833.

Kundzewicz, Z.W., Szamalek, K. and Kowalczak, P. (1999). The Great Flood of 1997 in Poland. *Hydrological Sciences Journal*, **44**(6), pp. 855-870.

Kunstmann, H., Kinzelbach, W. and Siegfried, T. (2002). Conditional first-order second-moment method and its application to the quantification of uncertainty in groundwater modelling. *Water Resources Research*, **38**(4), pp.6-1 to 6-15.

Langley, R.S. (2000). Unified Approach to Probabilistic and Possibilistic Analysis of Uncertain Systems. *Journal of Engineering Mechanics*, ASCE, **126**(11), pp. 1163-1172.

Lee, H.-L. and Mays, L.W. (1986). Hydraulic Uncertainty in Flood Levee Capacity. *Journal of Hydraulic Engineering*, ASCE, **112**(10), pp. 928-934.

Lettermann, A. (1991). Systerm-Theoretic Modelling in Surface Water Hydrology. Springer-Verlag, Berlin.

Leung, Y. (1982). Maximum entropy estimation with inexact information. *Fussy Set and Possibility Theory*, Yager, R.R. (ed), Pergamon Press, Oxford, pp. 32-37.

Levine, A. (1971). *Theory of Probability*. Addison-Wesley Publishing Company, Inc.

"Loire River." Encyclopaedia Britannica (2001). Encyclopaedia Britannica Premium Service. Retrieved 27 April 2001 http://www.britannica.com/eb/print?eu=49947.

Mantoglou, A. and Wilson, J.L. (1982). The turning bands method for simulation of random fields using line generation by a spectral method. *Water Resources Research*, **18**(5), pp. 1379-1394.

Margulis, S.A. and Entekhabi, D. (2001). Temporal disaggregation of satellite-derived monthly precipitation estimates and the resulting propagation of error in partitioning of water at the land surface. *Hydrology and Earth System Sciences*, **5**(1), pp. 27-38.

Markus, M., Tsai, C.W.-S. and Demissie, M. (2003). Uncertainty of weekly nitrate-nitrogen forecasts using artificial neural networks. *Journal of Environmental Engineering*, ASCE, **129**(3), pp. 267-274.

Maskey, S., Dibike, Y.B., Jonoski, A. and Solomatine, D.P. (2000a). Groundwater Model Approximation with Artificial Neural Network for Selecting Optimum Pumping Strategy for Plume Removal. In *Proceedings of the 2nd Joint Workshop on Artificial Intelligence Methods in Civil Engineering Applications*, Cottbus, Germany, 26-28 March 2000, pp. 67-77.

Maskey, S., Jonoski, A. and Solomatine, D.P. (2000b). Use of Global Optimisation Technique in Groundwater Pumping Strategy for Plume Removal. In *Groundwater Updates*, *Proceedings of International Symposium 2000 on Groundwater, IAHR*, Sato, K. and Iwasa, Y. (Eds.), 8-10 May 2000, Saitama, Japan, pp. 141-146.

Maskey, S. (2001). Uncertainty Analysis in Flood Forecasting and Warning System Using Expert Judgement and Fuzzy Set Theory. In *Safety & Reliability*, Zio, E., Demichela, M. and Piccinini, N. (eds.), pp. 1787-1794.

Maskey, S. and Guinot, V. (2002). Improved FOSM method for uncertainty analysis of a flood forecasting model. In Hydroinformatics 2002 Volume Two: Software Tools and Management Systems, Proceedings of the 5th International Conference of Hydroinformatics, 1-5 July 2002, Cardiff, UK, IWA Publishing, London, pp. 1331-1336.

Maskey, S., Guinot, V. and Delahaye, A. (2002a). Uncertainty Analysis of a Real Time Flood Forecasting Model. In *Advances in Hydro-Science and Engineering, Proceedings of the 5th International Conference "Hydro-Science and Engineering"*, 18-20 September 2002, Warsaw, Poland. Abstract p. 84, Full paper on CD.

Maskey, S., Jonoski, A. and Solomatine, D.P. (2002b). Groundwater remediation strategy using global optimization algorithms. *Journal of Water Resources Planning and Management*, ASCE, **128**(6), pp. 431-440.

Maskey, S. and Guinot, V. (2003). Improved First-Order Second Moment Method for Uncertaitny Estimation in Flood Forecasting. *Hydrological Sciences Journal*, **48**(2), pp. 183-196.

Maskey, S., Guinot, V. and Price, R.K. (2003a). Propagation of precipitation uncertainty through a flood forecasting model. In *Weather Rader Information and Distributed Hydrological Modelling* (Proceedings of the IAHS Symposium HS03, July 2003, Sapporo), IAHS Publ. No. **282**, pp. 93-100.

Maskey, S., Guinot, V. and Price, R.K. (2003b). Treatment of precipitation uncertainty in rainfall-runoff modelling: a fuzzy set approach. Under review by *Advances in Water Resources*.

Maskey, S. and Price, R.K. (2003a). Uncertainty Issues in flood forecasting. In *Flood Events: Are We Prepared?* (Proceedings of the OSIRIS Workshop, March 2003, Berlin), pp. 123-136.

Maskey, S. and Price, R.K. (2003b). Uncertainty assessment due to time series inputs using disaggregation for flood forecasting. *Proceedings of Netherlands Centre for River Studies (NCR) days 2003*, 6-8 November 2003, Roermond.

Maskey, S. (2004). Fuzzy-probabilistic risk based flood warning decision making. Will appear in *Proceedings of 6th International Conference on Hydro-science and Engineering 2004*, 30 May- 3 June, Brisbane, Australia.

Maskey, S. and Price, R.K. (2004). Assessment of uncertainty in flood forecasting using probabilistic and fuzzy approaches. Will appear in *Proceedings of 6th International Conference on Hydroinformatics 2004*, 21-24 June, Singapore.

McIntyre, N., Wheater, H. and Lees, M. (2002). Estimation and propagation of parametric uncertainty in environmental models. *Journal of Hydroinformatics*, **4**(3), pp. 177-198.

Melching, C.S. (1992). An improved first-order reliability approach for assessing uncertainties in hydrological modelling. *Journal of Hydrology*, 132, pp. 157-177.

Melching, C.S. (1995). Reliability Estimation. In *Computer Models of Watershed Hydrology*, Singh, V.P. (ed.), Water Resources Publications, Colorado, pp. 69-118.

Melching. C.S., Yen, B.C. and Wenzel Jr., H.G. (1991). Output reliability as guide for selection of rainfall-runoff models. *Journal of Water Resources Planning and Management*, ASCE, **117**(3), pp. 383-398.

Melching, C.S. and Yoon, C.G. (1996). Key Sources of Uncertainty in QUAL2E Model of Passaic River. *Journal of Water Resources Planning and Management*, ASCE, **122**(2), pp. 105-113.

Moore, R.E. (1962). Interval arithmetic and automatic error analysis in digital computing. PhD Thesis, Stanford University, California.

Moore, R.E. (1966). *Interval analysis*. Prentice-Hall, New York.

Moore, R.J. (2002). Aspects of uncertainty, reliability, and risk in flood forecasting systems incorporating weather radar. In *Risk, Reliability, Uncertainty, and Robustness of Water Resources Systems*, Eds. Bogardi, J.J. and Kundzewicz, Z.W., UNESCO, pp. 30-40.

Morgan, M.G. and Henrion, M. (1990). Uncertainty, A Guide to Dealing with Uncertainty in Qualitative Risk and Policy Analysis. Cambridge University Press, Cambridge, UK.

Nielsen, S.A. and Hansen, E. (1973). Numerical simulation of rainfall runoff process on a daily basis. *Nordic Hydrology*, **4**, pp. 171-190.

Ormsbee, L.E. (1989). Rainfall Disaggregation Model for Continuous Hydrologic Modelling. *Journal of Hydraulic Engineering*, ASCE, **115**(4), pp. 507-525.

Parker, D., Fordham, M. and Torterotot, J.-P. (1994). Real-time hazard management: Flood forecasting, warning and response. In *Floods Across Europe*, Penning-Rowsell, E.C. and Fordham, M. (eds.), Middlesex University Press, London, pp. 135-166.

Pawlowsky, M.A. (1995). Crossover Operations. In *Practical Handbook of Genetic Algorithms, Applications Vol. 1*, Chambers, L. (ed.). CRC Press, Inc., pp. 101-114.

Pedrycz, W. (1994). Why triangular membership functions? *Fuzzy Sets and Systems*, **64**, pp. 21-30.

Penning-Rowsell, E.C. and Peerbolte, B. (1994). Concepts, Policies and Research. In *Floods Across Europe*, Penning-Rowsell, E.C. and Fordham, M. (eds.), Middlesex University Press, London, pp. 1-17.

Price, R.K. (2000). Hydroinformatics for River Flood Management. In *Flood Issues in Contemporary Water Management, 2. Environmental Security*, Marsalek, J., Watt, W.E., Zeman, E. and Sieker, F. [eds.], **17**, NATO Science Series, Kluwer Academic Publishers, pp. 237-250.

Price, R.K. (2002). Hydroinformatics – Lecture 1. In *Proceedings of RBM 2002: Advanced study course on river basin modelling for flood risk mitigation*, 7-18 October 2002, The University of Birmingham, UK, pp. 5.1-5.19.

Radwan, M., Willems, P. and Berlamont, J. (2002). Sensitivity and uncertainty analysis for river water quality modelling. In *Hydroinformatics 2002: Proceedings of the Fifth International Conference on Hydroinformatics*, Cardiff, UK, IWA Publishing, London, pp. 482-487.

Reddy, R.K. and Haldar, A. (1992). A random-fuzzy reliability analysis of engineering systems. In *Analysis and Management of Uncertainty: Theory and Applications*. Ayyub, B.M., Gupta, M.M. and Kanal, L.N. (eds.), Elsevier Science Publishers, pp. 319-329.

Reed, D.W. (1984). *A review of British flood forecasting practice*. Report no. **90**. Institute of Hydrology, Wallingford, UK.

Revelli R. and Ridolfi, L. (2002). Fuzzy approach for analysis of pipe network. *Journal of Hydraulic Engineering*, ACSE, **128**(1), pp. 93-101.

Rodriguez-Iturbe, I. and Eagleson, P.S. (1987). Mathematical models of rainstorm events in space and time. *Water Resources Research*, **23**(1), pp. 181-190.

Romanowicz, R. and Beven, K. (1998). Dynamic Real-Time Prediction of Flood Inundation Probabilities. *Hydrological Sciences Journal*, **43**(2), pp. 181-196.

Romanowicz, R., Beven, K.J. and Tawn, J. (1996). Bayesian Calibration of Flood Inundation Models. In *Floodplain Processes*, M.G. Anderson, D.E. Walling and P.D. Bates (eds.), Wiley, Chichester, pp. 333-360.

Rosenblueth, E. (1975). Point estimates for probability moments. *Proceedings of the National Academy of Sciences* USA, **72**(10), pp. 3812-3814.

Rosenthal, U., Hart, P. and Bezuyen, M. (1998). Flood response and disaster management: a comparative perspective. In *Flood Response and Crisis Management in Western Europe*, Rosenthal, U. and Hart, P. (eds.), Springer-Verlag Berlin Heidelberg, pp.1-13.

Ross, T.J. (1995). *Fuzzy Logic with Engineering Applications*. McGraw-Hill, Inc., USA.

Salas, J.D. and Shin, H.-S. (1999). Uncertainty Analysis of Reservoir Sedimentation. *Journal of Hydraulic Engineering*, ASCE, **125**(4), 339-350.

Samuels, P.G. (1999). RIBAMOD, River Basin Modelling, Management and Flood Mitigation. Report SR **551**, HR Wallingford, UK.

Schulz, K. and Huwe, B. (1997). Water Flow Modelling in the Unsaturated Zone with Imprecise Parameters Using a Fuzzy Approach. *Journal of Hydrology*, **201**, pp. 211-229.

Schulz, K. and Huwe, B. (1999). Uncertainty and sensitivity analysis of water transport modelling in a layered soil profile using fuzzy set theory. *Journal of Hydroinformatics*, **1**(2), pp. 127-138.

Schulz, K., Huwe, B. and Peiffer, S. (1999). Parameter Uncertainty in Chemical Equilibrium Calculations Using a Fuzzy Set Theory. *Journal of Hydrology*, **217**, pp. 119-134.

Shafer, G. (1976). *A Mathematical Theory of Evidence*. Princeton University Press, Princeton, N.J.

Shamseldin, A.Y. (2002). Development of rainfall-runoff models. In *Proceedings of RBM 2002: Advanced study course on river basin modelling for flood risk mitigation*, 7-18 October 2002, The University of Birmingham, UK, pp. 9.1-9.27.

Shannon, C.E. (1948). A mathematical theory of communication. . *The Bell System Technical Journal*, **27**, pp. 379-423, 623-656.

Sherman, L.K. (1932). Streamflow from rainfall by the unit-graph method. Eng. New Rec., 108, pp. 501-505.

Sivakumar, B., Sorooshian, S., Gupta, H.V. and Gao, X. (2001). A chaotic approach to rainfall disaggregation. *Water Resources Research*, **37**(1), pp. 61-72.

Skaugen, T. (2002). A spatial disaggregation procedure for precipitation. *Hydrological Sciences Journal*, **47**(6), pp. 943-956.

Smith, K. and Ward, R. (1998). *FLOODS: Physical Processes and Human Impacts*. John Wiley and Sons.

Solomatine, D.P. (2002). Data-driven modelling: paradigm, methods, experiences. In *Hydroinformatics 2002: Proceedings of the Fifth International Conference on Hydroinformatics*, Cardiff, UK, IWA Publishing, London, pp. 757-763.

Song, Q. and Brown, L.C. (1990). DO Model Uncertainty with Correlated Inputs. *Journal of Environmental Engineering*, ASCE, **116**(6), pp. 1164-1180.

Sugawara, M., Watanabe, I., Ozaki, E. and Katsuyame, Y. (1983). *Reference manual for the TANK model*. National Research Centre for Disaster Prevention, Tokyo.

Sundararajan, C.R. (1994). Uncertainties in piping frequency analysis. *Fuzzy Sets and Systems*, **66**, pp. 283-292.

Sundararajan, C.R. (1998). Fuzzy Models for Uncertainty Assessment in Engineering Analysis. In *Uncertainty Modelling and Analysis in Civil Engineering*, Ayyub, B.M. [ed.], CRS Press, pp. 43-53.

Szamalek, K. (2000). The great flood of 1997 in Poland: the truth and myth. In *Flood Issues in Contemporary Water Management, 2. Environmental Security*, Marsalek, J., Watt, W.E., Zeman, E. and Sieker, F. [eds.], **17**, NATO Science Series, Kluwer Academic Publishers, pp. 67-83.

Tarboton, D.G., Sharma, A. and Lall, U. (1998). Disaggregation procedures for stochastic hydrology based on nonparametric density estimation. *Water Resources Research*, **34**(1), pp. 107-119.

Terano, T., Asai, K. and Sugeno, M. (1992). *Fuzzy Systems Theory and Its Applications*. Academic Press, Inc, London, UK.

Tsoukalas, L.H. and Uhrig, R. E. (1997). *Fuzzy and Neural approaches in Engineering*. A Wiley-Interscience Publication, John Wiley & Sons, Inc., New York.

Tung, Y.-K. and Mays, L.W. (1981). Risk Models for Flood Levee Design. *Water Resources Research*, **17**(4), pp. 833-842.

Tyagi, A. and Haan, C.T. (2001). Uncertainty analysis using corrected first-order approximation method. *Water Resources Research*, **37**(6), pp. 1847-1858.

USACE (1998). HEC-1 Flood Hydrograph Package User's Manual. *Hydrologic Engineering Centre*, Davis, CA.

USACE (2000). Hydrologic Modelling System HEC-HMS Technical Reference Manual. *Hydrologic Engineering Centre*, Davis, CA.

USACE (2001). Hydrologic Modelling System HEC-HMS User's Manual. *Hydrologic Engineering Centre*, Davis, CA.

Van der Klis, H. (2003). *Uncertainty analysis applied to numerical models of river bed morphology*. PhD Thesis, Technical University Delft, The Netherlands.

Van Gelder, P.H.A.J.M. (2000). *Statistical methods for the risk-based design of civil structures*. PhD Thesis, Technology University Delft, The Netherlands.

Voortman, H.G. (2002). *Risk-based design of large-scale flood defence systems*. PhD. Thesis, Technical University Delft, The Netherlands.

Vrijling, J.K. (1993). Development in Probabilistic Design of Flood Defences in the Netherlands. In *Reliability and Uncertainty Analysis in Hydraulic Design*, Yen, B.C. and Tun, Y.-K., ASCE, pp. 133-178.

Vrijling, J.K., Van Hengel, W. and Houben, R.J. (1998). Acceptable risk as a basis for design. *Reliability Engineering & System Safety*, **59**(1), pp. 141-150.

Ward, R. (1978). *Floods: A geographical perspective*. The Macmillan Press Ltd., London.

Warwick, J.J. and Wilson, J.S. (1990). Estimating Uncertainty of Stormwater Runoff Computations. *Journal of Water Resources Planning and Management*, ASCE, **116**(2), pp. 187-204.

Wetherill, G.B. (1981). *Intermediate Statistical Methods*. Chapman and Hall, New York.

Wonneberger, S. (1994). Generalization of an invertible mapping between probability and possibility. *Fuzzy Sets and Systems*, **64** (1994), 229-240.

Yager, R.R. (1986). A Characterization of the Extension Principle. In *Fuzzy Sets and Systems*, **18**, pp. 205-217.

Yang, G., Reinstein, L.E., Pai, S., Xu, Z. and Carroll, D.L. (1998). A new genetic algorithm technique in optimization of permanent prostate implants. *Medical Physics*, **25**(12), pp. 2308-1315.

Yu, P.-S. and Tseng, T.-Y. (1996). A model to forecast flow with uncertainty analysis. *Hydrological Sciences Journal*, **41**(3), pp. 327-343.

Yu, P.-S., Yang, T.-C. and Chen, S.-J. (2001). Comparison of uncertainty analysis methods for a distributed rainfall-runoff models. *Journal of Hydrology*, **244**, pp. 43-59.

Zadeh, L.A. (1965). Fuzzy sets. *Information and Control*, **8**, pp. 338-353.

Zadeh, L.A. (1968). Probability Measures of Fuzzy Events. *Journal of Mathematical Analysis and Applications*, **22**, pp. 421-427.

Zadeh, L.A. (1975). The Concept of a Linguistic Variable and its Application to Approximate Reasoning, Part I. *Information Sciences*, **8**, pp. 199-249.

Zadeh, L.A. (1978). Fuzzy Sets as a Basis for Theory of Possibility. *Fuzzy Sets and Systems*, **1**, pp. 3-28.

Zadeh, L.A. (1984). Fuzzy Probabilities. *Information Processing and Management*, **20**(3), pp. 363-372.

Zadeh, L.A. (1995). Discussion: Probability Theory and Fuzzy Logic Are Complementary Rather than Competitive. *Technometrics*, **37**(3), pp. 271-276.

Zimmermann, H.-J. (1991). *Fuzzy set theory and its applications*, Kluwer Academic Publishers, Dordrecht.

Zimmermann, H.-J. (1997a). A Fresh Perspective on Uncertainty Modelling: Uncertainty vs. Modelling. In *Uncertainty Analysis in Engineering and Sciences: Fuzzy Logic, Statistics, and Neural Network Approach*, Ayyub, B.M. and Gupta, M.M. [eds.], Kluwer Academic Publisher, pp. 353-364.

Zimmermann, H.-J. (1997b). Uncertainty Modelling and Fuzzy Sets. In *Uncertainty: Models and Measures*, Natke, H. G. and Ben-Haim, Y. (eds.), Mathematical Research, **99**, Akademie Verlag, pp. 84-100.

Samenvatting

MODELLERING VAN ONZEKERHEID IN SYSTEMEN VOOR HOOGWATERVOORSPELLING

Onzekerheid is een gebruikelijk onderdeel van het dagelijkse leven. Onder vrijwel alle omstandigheden bevinden we ons in een toestand van onzekerheid. Het onderwerp van dit onderzoek is de onzekerheid in systemen van hoogwatervoorspelling. Zoals alle natuurlijke gevaren is hoogwater een inherent onzeker fenomeen. De onzekerheid in hoogwatervoorspelling is het resultaat van de onzekerheid in de neerslagverwachting en andere invoer, in modelparameters, modelstructuur, enzovoort. Ondanks de toenemende vooruitgang in de ontwikkeling van modellen en technieken voor hoogwatervoorspelling, blijft onzekerheid in de actuele hoogwatervoorspelling onvermijdelijk. Daarom is het van belang dat het bestaan van onzekerheid in hoogwatervoorspellingen wordt onderkend en op de juiste wijze wordt geschat. Het verbergen van onzekerheid kan de illusie van zekerheid creëren en de consequenties daarvan kunnen aanzienlijk zijn. De grootste voordelen van het schatten van onzekerheid in hoogwatervoorspelling is dat dit een rationele basis geeft voor hoogwaterwaarschuwing (op risico gebaseerde waarschuwing) en dat het in potentie economische voordelen biedt die algemeen samenhangen met hoogwatervoorspelling en waarschuwingssystemen. Zelfs zonder risicogebaseerde procedures voor hoogwaterwaarschuwing geeft het kwantificeren van de onzekerheid extra informatie over de voorspelling en helpt daarmee beslissers op de juiste gronden hun eigen beoordeling te maken.

Het doel van dit onderzoek is om een raamwerk, en gereedschappen en technieken te ontwikkelen voor het modelleren van onzekerheid in systemen voor hoogwatervoorspelling. In het onderzoek worden methoden van onzekerheidsschatting toegepast, die zijn gebaseerd op waarschijnlijkheidstheorie en de theorie van fuzzy sets, op het probleem van hoogwatervoorspelling. De uitkomsten van het onderzoek kunnen als volgt worden samengevat:

1. Inventarisatie van verschillende typen modellen voor hoogwatervoorspelling en hun beoordeling met betrekking tot onzekerheidsschatting.

2. Inventarisatie van de weergave van onzekerheid en de theorieën en methoden van modelleren, in het bijzonder ten aanzien van hoogwatervoorspelling.

3. Ontwikkeling van een methodologie waarbij gebruik wordt gemaakt van temporele disaggregatie van onzekerheidsschatting in modeluitvoer als gevolg van onzekerheid in tijdreeksinvoer.

4. Ontwikkeling van een verbeterde eerste orde secondair moment methode (IFOSM) voor de schatting van onzekerheid waarbij gebruik wordt gemaakt van een tweede orde reconstructie van de modelfunctie.

5. Ontwikkeling van een kwalitatieve onzekerheidsschaal waarbij gebruik wordt gemaakt van best-case en worst-case scenario's voor de interpretatie van resultaten van onzekerheidsanalyse, gegenereerd met een op expert-judgement gebaseerde kwalitatieve methode.

6. Verkenning van hybride technieken van onzekerheidsmodellering en kans-waarschijnlijkheidstranformaties.

Er zijn verschillende modellen voor hoogwatervoorspelling geïnventariseerd. Globaal kunnen deze worden geclassificeerd als: (i) op fysica gebaseerd, (ii) conceptueel en empirisch, en (iii) op gegevens gebaseerd. Onzekerheidsanalyse is een breed geaccepteerde procedure in de eerste twee categorieën van modellen, alhoewel veel operationele voorspellingssystemen voor hoogwater geen componenten voor onzekerheidsanalyse bevatten. Normaliter leidt toenemende complexiteit van het model tot een reductie van de onzekerheid van de modeluitvoer, als gevolg van de structuur van het model. Op deze wijze zou de modelonzekerheid moeten reduceren indien we de stap maken van empirische, via conceptuele, naar op fysica gebaseerde modellen. Echter, modelonzekerheid is slechts een deel van de totale onzekerheid. Indien er grote onzekerheid bestaat in de invoer en als de parameters niet met hoge nauwkeurigheid kunnen worden bepaald, garandeert het gebruik van complexe modellen niet dat de modelresultaten minder onzeker zijn. Sommige gegevensgestuurde technieken, zoals op fuzzy regels gebaseerde systemen en fuzzy regressie, werken met onnauwkeurige gegevens en verwerken het onzekerheidsconcept in de modellering. Deze modellen echter, kennen geen flexibiliteit in het gebruik van onzekerheidsmethoden die op andere theorieën zijn gebaseerd (bijvoorbeeld de populaire probabilistische theorie). In het geval het model is gebaseerd op een kunstmatige neuraal netwerk (ANN) (de meest populaire van de op gegevens gebaseerde technieken), wordt de modelperformance veelal uitgedrukt in termen van het verschil tussen waargenomen en door het model voorspelde resultaten, waarbij maten zoals de kleinste kwadraten methode (RMSE) worden toegepast. Er is meer onderzoek nodig naar de toepassing van de meer gebruikelijke technieken van onzekerheidsanalyse, zoals die welke zijn gebaseerd op probabilistiek en fuzzy set theorie, op ANN modellen.

Het literatuuronderzoek geeft aan dat de theorie van de probabilistiek en de theorie van fuzzy sets (inclusief fuzzy maatregelen en waarschijnlijkheidstheorie) de twee meest gebruikte theorieën zijn voor de weergave van onzekerheid. De theorie van de probabilistiek veronderstelt dat onzekerheid grotendeels het gevolg is van het random proces, terwijl de fuzzy set theorie veronderstelt dat dit het gevolg is van vaagheid (fuzzyness) en onnauwkeurigheid. In hoogwatervoorspelling is de toepassing van theorieën, anders dan de theorie van de probabilistiek, tot nog toe evenwel van geen betekenis. Er wordt beargumenteerd dat de twee theorieën eerder als complementair moeten worden behandeld, dan als elkaar beconcurrerend. In dit onderzoek zijn de Monte Carlo en eerste-orde tweede moment (FOSM) methode gebruikt als standaard technieken voor de voortplanting van onzekerheid in de probabilistische aanpak. Tegelijkertijd breidt dit onderzoek de toepassing van de fuzzy set theorie uit voor het modelleren van onzekerheid in hoogwatervoorspelling. In het bijzonder zijn twee methoden verkend die zijn gebaseerd op de fuzzy set theorie: (i) de fuzzy Extension Principle, en (ii) de op expert judgement gebaseerde kwalitatieve methode.

Een belangrijke bijdrage van dit onderzoek betreft de ontwikkeling van een methodologie die gebruik maakt van temporele de-aggregatie voor onzekerheidsschatting in modelresultaten als gevolg van onzekerheid in tijdreeksinvoer. Deze methodologie kan worden geïmplementeerd in raamwerken van zowel de Monte Carlo methode als van de fuzzy Extension Principle, en kan worden toegepast voor ieder type deterministisch neerslag-afvoer model. Deze methodologie vereist op expliciete wijze dat de onzekerheid in de tijdreeksinvoer wordt gerepresenteerd door kansverdelingen in het geval van de probabilistische benadering en door membership functies in geval van de fuzzy benadering. Deze laatste benadering is in het bijzonder bruikbaar als de neerslagverwachtingen niet-probabilistisch zijn.

Het feit dat de populaire FOSM methode gebruik maakt van linearisatie van de modelfunctie geeft aanleiding tot beperkingen van de methode. Als onderdeel van dit onderzoek wordt een verbeterde FOSM methode voorgesteld (IFOSM) waarbij gebruik wordt gemaakt van de tweede orde reconstructie van de functie die moet worden gemodelleerd. De IFOSM methode behoudt de eenvoud en de beperkte rekenkracht van de FOSM methode en heeft een bijzonder voordeel indien de gemiddelde waarde van de invoervariabele overeenkomt met het maximum/minimum van de functie, of waarden in regionen waarbij helling van de functie redelijk beperkt is in vergelijking met de effecten van de kromming (niet-lineariteit).

Het onderzoek laat ook de toepassing van een kwalitatieve methode zien voor de schatting van onzekerheid in hoogwatervoorspelling. De kwalitatieve methode is gebaseerd op expert judgement en de fuzzy set theorie. Het voordeel van deze methode is dat het de mogelijkheid biedt om te werken met onzekerheid als gevolg van verschillende te onderkennen bronnen, zonder noemenswaardige toename in de rekentijd. De methode heeft dezelfde mathematische structuur als die van de FOSM methode, echter de evaluaties van de kwaliteit en het belang (vergelijkbaar met respectievelijk de variantie en de gevoeligheid van de FOSM methode) zijn volledig gebaseerd op expert judgement. Dientengevolge is dit een zeer benaderende en tevens meer holistische methode. Eén van de belangrijke onderwerpen bij deze methode betreft de interpretatie van de resultaten. Gedurende het onderzoek is een kwalitatieve onzekerheidsschaal (Qualitiative Uncertainty Scale) ontwikkeld op basis waarvan de resultaten van de op expert judgement gebaseerde methode, kwalitatief kunnen worden geclassificeerd met het gebruik van linguïstische variabelen zoals: 'kleine onzekerheid', 'gematigde onzekerheid', 'grote onzekerheid', etc. De afleiding van de onzekerheidsschaal is gebaseerd op het concept van best-case en worst-case scenario's.

Het onderzoek behandelt ook de mogelijkheid van een hybride techniek van onzekerheidsmodellering, waarbij de gecombineerde toepassing van zowel de probabilistische als de waarschijnlijkheid of fuzzy benadering worden gebruikt. Er zijn ten minste twee typen situaties waarbij de hybride techniek bruikbaar is, in dit onderzoek aangegeven als 'Type I' en 'Type II' problemen. Het Type I probleem komt voor als een onzekere variabele componenten van randomness (willekeur) en fuzziness (vaagheid) in zich heeft. Het Type II probleem heeft betrekking op de situatie waarbij er twee sets van onzekerheidsparameters zijn: één gerepresenteerd door de kansverdeling en de andere weergegeven door de waarschijnlijkheidsverdeling of membership functie. De concepten van fuzzy probabilistiek en fuzzy random variabelen kunnen worden gebruikt in het Type I probleem, terwijl een oplossing van het Type II probleem ook mogelijk is als er een transformatie tussen de kansverdeling en de waarschijnlijkheid kan worden vastgesteld.

Het grootste probleem in de modellering van onzekerheid van het Type II probleem is het resultaat van de verschillen in toepassing van random-random (R-R) en fuzzy-fuzzy (F-F) variabelen. Een deel van dit onderzoek richt zich op het verkennen van de verschillen en overeenkomsten tussen R-R en F-F variabelen en het onderzoeken van de afleiding van de kans-waarschijnlijkheid (of fuzzy) transformatie die rekening houdt met deze verschillen. Het onderzoek laat zien dat de rekenkundige bewerkingen tussen twee fuzzy variabelen door de fuzzy Extension Principle, gebruikmakend van een methode die is aangeduid met 'α-cut', voor enkele specifieke condities vergelijkbaar zijn met de overeenkomstige bewerkingen tussen twee functioneel afhankelijke random variabelen. Er wordt ook een alternatieve methode gegeven voor de evaluatie van de Extension Principle voor niet-monotone functies, zonder toepassing van de α-cut methode. Hoewel de directe implicatie van deze vinding beperkt is, vormt deze zeker een belangrijke basis voor verder onderzoek in de kans-waarschijnlijkheid transformatie en hybride benaderingen in de onzekerheidsmodellering.

De praktijktoepassingen die zijn behandeld in dit onderzoek hebben betrekking op de hoogwater voorspellingsmodellen van het Klodzko afvoergebied in Polen en van de Loire rivier in Frankrijk. De methodologie van onzekerheidsschatting, waarbij gebruik wordt gemaakt van temporele de-aggregatie, is toegepast op het Klodzko model om de onzekerheid te bepalen in de voorspelde hoogwatersituaties die het gevolg zijn van onzekerheid in de neerslagtijdreeks. De toepassing vindt plaats in het raamwerk van het fuzzy Extension Principle en wordt ondersteund met genetische algoritmen (GA's) voor de bepaling van de minimale en maximale waarden van de modelfunctie. De resultaten worden geïllustreerd met twee versies van GA's, namelijk de normale (conventionele) GA en een micro GA. Het Loire model is gebruikt voor de toepassing van het FOSM, IFOSM en de op expert judgement gebaseerde kwalitatieve methoden van onzekerheidsschatting in de modelvoorspellingen, die het resultaat zijn van onzekerheid in verschillende parameters. In deze toepassing is de Monte Carlo methode gebruikt als standaard methode ter vergelijking van de resultaten van de FOSM en de IFOSM methoden.

Shreedhar Maskey

Delft, 2004

Acknowledgements

First of all, I am very grateful to my promoter Prof. Roland K. Price for offering me a PhD opportunity at UNESCO-IHE and continuously guiding me through my long endeavour of PhD research. Without his kindness, support and precious thoughts I wouldn't have passed through the journey, which is rocky, undulating and strenuous. I am also very grateful to Dr. Vincent Guinot for his persistent supervision and guidance throughout my PhD endeavor. I am very fortunate to have him as my supervisor, who assumed full responsibility not only when he was at UNESCO-IHE but also after he left for the University of Montpellier. I have learnt and benefited a lot from his precious suggestions and ideas. I also received occasional supervision from Prof. Jean Cunge during the early, yet crucial, stage of my research for which I am very grateful. I have benefited very much not only from his vast knowledge and experience in flood modelling but also from his very pragmatic advice.

During my research, I am privileged to meet and discuss my research with Prof. Andras Bardossy, University of Stuttgart, Prof. J.K. Vrijling, Technical University Delft, Prof. J. Philip O'Kane, National University of Ireland, Prof. Mike Hall, UNESCO-IHE and Prof. Peter Goodwin, University of Idaho. I am very grateful to all of them for their invaluable ideas and advices.

Most of this research was sponsored by an EU project "Operational Solutions for the Management of Inundation Risks in the Information Society" (OSIRIS), contract IST-1999-11598. I enjoyed very much working with the OSIRIS partners, all of whom were very friendly and helpful. I am particularly thankful to Roman Konieczny and Pawel Madej, Institute of Meteorology and Water Management, Krakow (Poland) and Arnaud Delahaye, DIREN Centre, Orleans (France) for providing data about the Klodzko and Loire catchments.

It has mostly been challenging yet fun to work within the Department of Hydroinformatics and Knowledge Management at UNESCO-IHE. All the staff members have been very friendly and always supportive; my thanks go to all of them. I have worked closely with Dr. Arnold Lobbrecht, first as a researcher in a project and later for Real Time Control lectures. I enjoyed and learnt a lot from working with him. He also prepared the Dutch translation of the summary of this thesis. I am very grateful to him. I also received much help from Dr. Dimitri Solomatine for which I am very grateful. I also appreciate his personal notes to my family. I had to take financial issues several times to Ir. Jan Luijendijk, the head of the department. I am very thankful to him for he always took these issues positively. I am also thankful to all the staff of UNESCO-IHE in general and the ladies in Student Affairs in particular, who provided me with a lot of cooperation, directly or indirectly.

Perhaps, no one understands better the suffering of a PhD fellow than a PhD fellow. Therefore, my thanks are also to the PhD fellows at UNESCO-IHE with whom I shared my sufferings, and also my joys.

For my MSc and PhD studies at UNESCO-IHE, I remained away from my home country for more than 6 years. I certainly missed not only many of my friends and loved ones, but also the culture, the carnivals, the hills and mountains and all the natural beauties. But, my thanks go to all the Nepali friends at IHE and in The Netherlands, whose presence never let me become desperate because of this sentiment.

Last but certainly not least, the credit for all my achievements goes to my wife Neeru and daughter Shreta. Their patience, support and company are the key factors for my success. Many thanks to you both.

About the Author

Shreedhar Maskey was born on 24 December 1966 in Charikot, in the district Dolkha, Nepal. In 1990, he graduated in Civil Engineering from Tribhuvan University in Nepal. Since 1991 he served in the same university as a lecturer until he joined UNESCO-IHE (then IHE-Delft) in 1997 to participate in the Master's Programme in Hydroinformatics. He received the Master of Science degree with distinction in hydroinformatics in April 1999. Since April 1999 he has been at UNESCO-IHE as a PhD research fellow. During this period he also worked for the COW[1] project (Apr. 1999 – Dec. 1999) and the OSIRIS[2] project (Jan. 2000 – Mar. 2003) and assisted in various ways in the Hydroinformatics Master's Programme.

He spent six months during 1998/1999 at HR Wallingford, UK as a visiting academic where he carried out research for his MSc dissertation. As a lecturer at Tribhuvan University he was teaching in particular engineering mechanics and structures. Besides lecturing, he also worked as an Assistant Campus Chief at the Pulchowk Engineering Campus of the university for one year in 1996/1997. During the period from 1991 to1997, he also worked on a number of civil engineering projects, mainly for structural analysis and design.

Mr Maskey received "Groundwater Science and Engineering Award" in the IAHR symposium on groundwater held between 8-10 May 2000 in Saitama, Japan. He has published a number of technical papers in international journals and conference proceedings. He is a member of the International Association of Hydrological Sciences (IAHS), the Nepal Engineers' Association (NEA) and the Computer Association of Nepal (CAN) and an International Student member of the American Society of Civil Engineers (ASCE) (2001-2003).

During his school days, the author had to walk to and from school for three quarters of an hour, each way. Several small streams pass through the way from his home to school, some of which used to swell unusually following heavy monsoon rains. Occasionally, during summer days, he and his colleagues had either to wait for hours until the swollen streams subsided or had to run fast in the heavy rain before they become potentially threatening. Until Mr. Maskey undertook his PhD project, he had no clue that one day he would be studying so closely the natural phenomenon that he personally witnessed a number of times during his boyhood.

---

[1] *Computer ondersteund water beheer* (In Dutch: Computer Assisted Water Management), a Dutch Water Authority funded project.
[2] Operational Solutions for the Management of Inundation Risks in the Information Society, an EC funded project carried out by 13 different organisations from five EU nations (France, Germany, Italy, Netherlands and Poland).

For Product Safety Concerns and Information please contact our EU
representative GPSR@taylorandfrancis.com Taylor & Francis Verlag GmbH,
Kaufingerstraße 24, 80331 München, Germany

Printed and bound by CPI Group (UK) Ltd, Croydon, CR0 4YY
01/05/2025
01858491-0001